BRAIN BENDERS ADVANCED

Challenging Puzzles and Games for Math and Language Arts

Credits:
Editor: Julie Kirsch
Layout Design: Chasity Rice
Cover Design: Chasity Rice
Cover Illustration: Bill Neville, Mike Duggins

ISBN 978-1-60022-314-3

TABLE OF CONTENTS

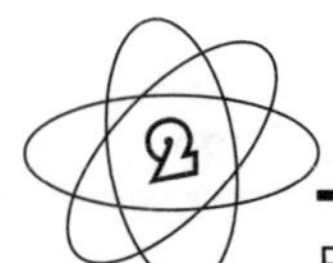

HELP PROFESSOR WHIZ

1. Professor Whiz had nine balls numbered from 1 to 9. The professor wanted to arrange the balls in the shape of a triangle so that the four numbers along each side of the triangle had the same sum. Write the numbers in the balls to show how he could do this.

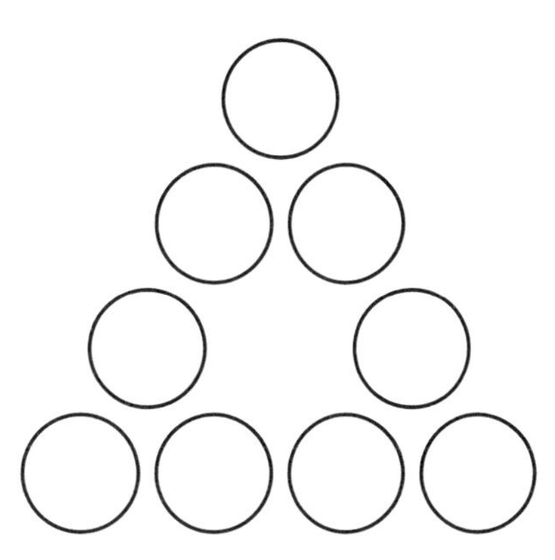

2. One of the professor's students gave him a ball with the number *0* on it. Professor Whiz decided to arrange the 10 balls to make another triangle. He still wanted the four numbers along each side to have the same sum. Write the numbers in the balls to show how he could do this, but do not put the 0 in the middle ball.

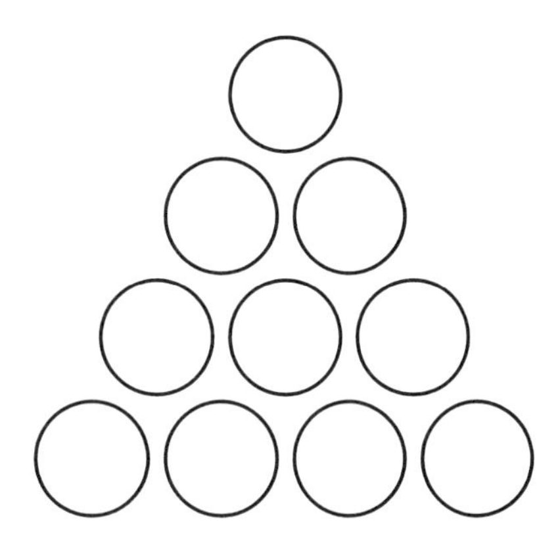

PYRAMID DOLLARS

1. Write the remaining dollar amounts on the pyramid to make a total of $10,000. Use only the number *5*. **HINT:** Use only whole dollar amounts.

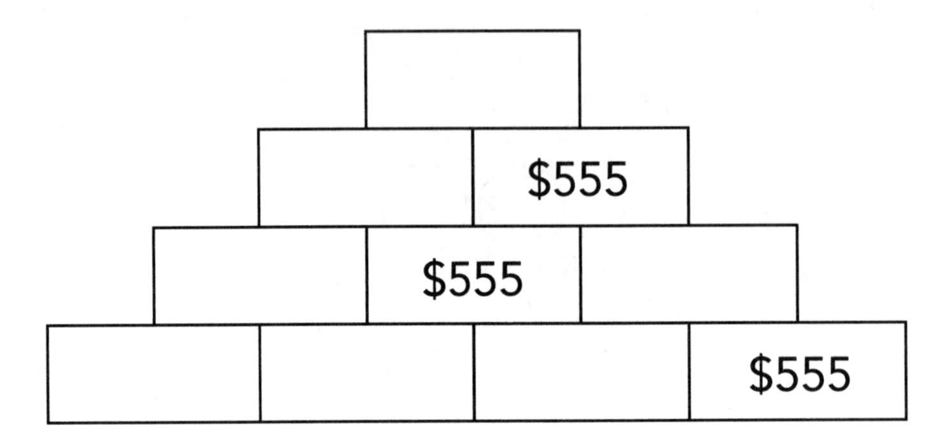

2. Write the remaining dollar amounts on the pyramid to make a total of $10,000. Use only the number *4*. **HINT:** Use only whole dollar amounts.

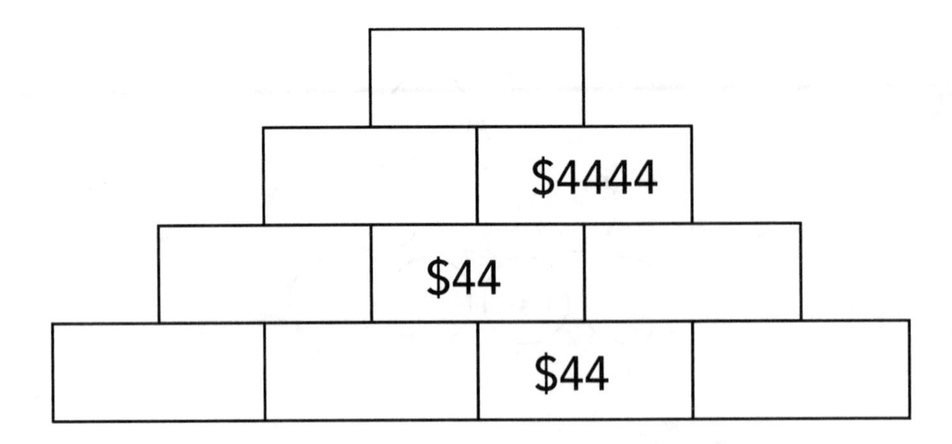

MYSTERY DIGITS

One number in each equation has been replaced with the letter *x*. Write the number for *x* that makes each equation true on the line provided.

1. $\frac{x}{8} + \frac{x}{8} = \frac{x}{x}$

   x = ______

2. $\frac{2}{x} + \frac{2}{x} = 2$

   x = ______

3. $x + \frac{x}{x} = 4$

   x = ______

4. $\frac{x}{x} - \frac{x}{9} = \frac{x}{18}$

   x = ______

5. $\frac{x}{9} - \frac{1}{x} = 0$

   x = ______

6. $xx + \frac{x}{x} = 100$

   x = ______

7. $\frac{(x)(x) - x}{x} = 3$

   x = ______

8. $\frac{(x)(x) + x}{x} = 8$

   x = ______

9. $\frac{xx}{x} - x = 10$

   x = ______

10. $\frac{xx}{x} - 6 = x$

    x = ______

AN AMAZING RECORD

Follow the steps to learn about an amazing record.

Step 1: Write the number *2,144,448,000* on a separate sheet of paper.

Step 2: Divide the number by the product of 12 and 5.

What do you get? ______________________

Step 3: Divide the quotient by the sum of 33 and 27.

What do you get? ______________________

Step 4: Divide that quotient by the quotient of 48 and 2.

What do you get? ______________________

Step 5: Finally, divide that quotient by the difference between 454 and 89.

What do you get? ______________________

The answer in Step 5 represents the number of years that a person had the hiccups!

Bonus! What unit of measurement does *2,144,448,000* represent?

____

COUSIN TALK

1. Cory and Keith are cousins who live exactly 36 miles apart. The boys decide to meet halfway between their houses. At the same time, they start riding their bikes toward each other. Each boy is riding 6 miles per hour.

   Keith has a dog that starts running the instant that Keith starts biking. The dog runs back and forth at 18 miles per hour between the two cousins' houses until the cousins meet. How far does the dog travel?

2. Kyle, Gladys, and Robert are cousins. All three cousins have different heights. Kyle is 14 inches taller than Gladys. The difference between Kyle and Robert is 2 inches less than the difference between Robert and Gladys. At 6'6", Kyle is the tallest. How tall are Robert and Gladys?

3. Jim wants to buy his cousin a guitar as a birthday gift, but the guitar costs $625. Jim decides to wait until the guitar is on sale. Luckily, the store has a sale the following week, and Jim sees that the guitar then costs $400. Unfortunately, that is still too expensive. The following week, he returns to the store and finds the guitar priced at $256. Jim decides to wait one more week. The next week, Jim sees that the price has been reduced again, at the same rate that it was reduced in each of the previous weeks. Jim buys the guitar. How much does he pay for it?

MR. KERR'S DOGS

1. Mr. Kerr is a math teacher. One day, he asked his students to guess the ages of his three dogs. He said, "I have had each dog for at least 2 years. If you multiply their ages, you get 36."

   There was no response. The class looked puzzled.

   Mr. Kerr continued, "If you add their ages, you get an odd number."

   The class still looked puzzled.

   Mr. Kerr said, "This is my final clue. The oldest dog was a gift from my mother. The youngest dog was a gift from my wife."

   Suddenly, a student called out, "I know the ages, Mr. Kerr!"

What are the ages of Mr. Kerr's dogs? ______________________________

Explain how you got your answer. ______________________________

____

____

2. Mr. Kerr has three dogs—a German shepherd, an Irish setter, and a Scottish terrier. Each dog has a mat of a different color to sleep on—a blue mat, a yellow mat, or a red mat. Use the clues to find each dog's name, breed, and mat color.

   - The German shepherd does not have the yellow mat.
   - Bandit has the blue mat.
   - The Irish Setter's name is either Jericho or Pepper.
   - Jericho does not have the red mat.
   - The Scottish terrier loves its red mat.

Write the name of each dog, its breed, and its mat color. ______________________________

____

____

____

MS. BAKER'S BROWNIES

Ms. Baker is famous for her delicious brownies. She usually bakes each batch in her 8 x 8-inch pan. One batch makes 16 brownies that are all the same size.

Ms. Baker needs to bake 72 brownies for a family reunion. She decides to use her 12 x 12-inch pan to bake the brownies instead. That way, she can bake more brownies at a time. Ms. Baker wants the brownies to be the same size as the brownies she makes in her 8 x 8-inch pan. How many batches will she need to bake?

Write your strategy and the solution. ______________________________

____

____

____

____

____

____

____

____

____

MULTIPLICATION PUZZLER

Replace each letter with a number to make the multiplication problem in the box true. Each letter represents a different digit. The questions will help you narrow your search.

$$\begin{array}{r} ABCDE \\ \times\ 4 \\ \hline EDCBA \end{array}$$

1. How many digits are in the number being multiplied by 4? ______________________

2. How many digits are in the product? ______________________

3. What is the smallest number that *ABCDE* can be? ______________________

   Why? ______________________

4. What is the largest number that *ABCDE* can be? ______________________

   Why? ______________________

5. What is the solution? Use the space below to solve the problem.

   ____

HOT DOG SURVEY

Mrs. Cook runs the school cafeteria. She took a survey of 100 students to find out how they ate their hot dogs. Here are the results:

- 40 students put at least relish on their hot dogs.
- 55 students put at least mustard on their hot dogs.
- 65 students put at least ketchup on their hot dogs.
- 15 students put on at least relish and mustard on their hot dogs.
- 35 students put on at least ketchup and mustard on their hot dogs.
- 20 students put on at least relish and ketchup on their hot dogs.
- 10 students put relish, mustard, and ketchup on their hot dogs.

What percentage of the students put relish and/or mustard on their hot dogs?

AT THE RESTAURANT

1. Six people met at a restaurant for a business luncheon. They all shook hands with one another before they sat down. How many handshakes were there in all?

2. The restaurant menu included a salad with lunch. Customers could choose between a lettuce salad or a spinach salad. They could also choose a topping of cheese, bacon bits, or sesame seeds. The choices for dressing were ranch, Italian, or blue cheese. How many different salad combinations were possible?

3. Amy, Brian, Carl, and Dee were planning to sit at a circular table. They were trying to decide where each person should sit. How many different ways could they sit around the table? **HINT:** Because the table is circular, arrangements such as the two shown are considered identical.

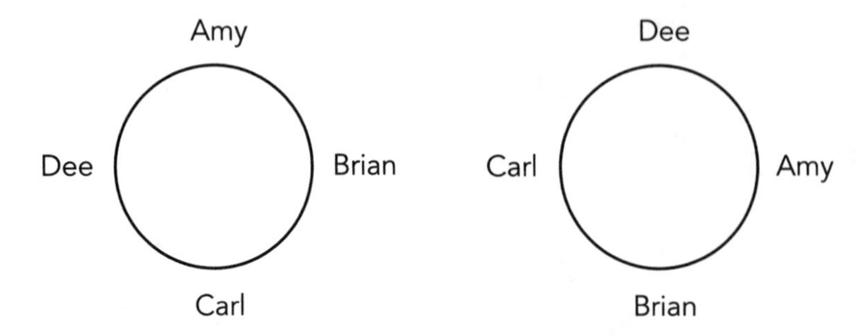

LINE THEM UP

1. Write the numbers *5*, *6*, *7*, and *8* in the boxes so that each row, column, and 4-square diagonal contains all of the numbers.

5	6	7	8

2. Now, write the numbers *5*, *6*, *7*, *8*, and *9* in the boxes so that each row, column, and 5-square diagonal contains all of the numbers.

5	6	7	8	9

LETTER GRID

Write the letters in the grid using the clues.

1. M is in the same column as P and S.
2. Q is directly above O and directly to the left of M.
3. R is directly to the right of P and directly above T.
4. O is directly to the left of S.
5. L is in the same row as P and R.
6. N is in the same column as R and T.

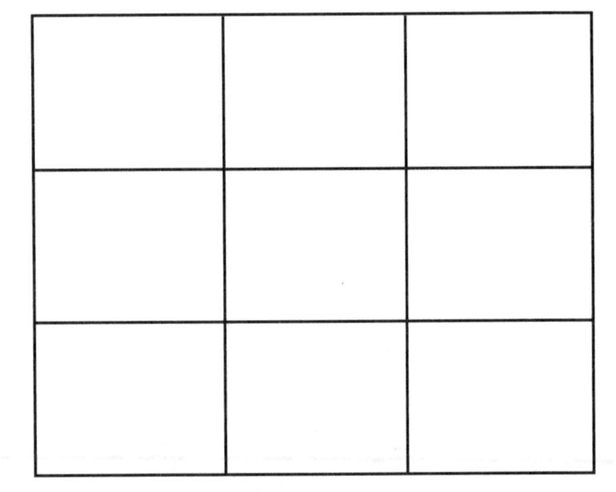

A CLASS PARTY

Mr. Garcia and his students are planning a class party. Seven students have volunteered to bring either a snack or a drink. The students are Ashley, Brad, David, Emma, Jill, Mike, and Rick. Use the clues to find what each student has volunteered to bring.

1. David and Brad will not bring nuts or popcorn.
2. Emma will not bring cookies.
3. No boy volunteered to bring juice.
4. Ashley and Jill said they would bring fruit. One will bring grapes and the other will bring cherries.
5. One of the boys will bring popcorn, but it won't be Rick.
6. Jill and Brad each said that they would bring something that starts with the letter *c*.

	nuts	popcorn	cookies	juice	grapes	cherries	punch
Ashley							
Brad							
David							
Emma							
Jill							
Mike							
Rick							

MONEY BAGS

Sherri's grandpa got eight paper bags. He put a five-dollar bill in three of the bags and a blank piece of paper in each of the remaining five bags. Then, he labeled the bags from A to H, sealed them with tape, and placed them on a table.

He told Sherri that she could have all of the five-dollar bills in the bags if she could figure out which bags held the money. He gave her a list with the following clues:

1. ACE
2. BDG
3. ACH
4. DFG
5. BCH
6. BEF
7. ADG
8. ABH
9. CEF
10. AGH
11. AEF
12. BGH

Grandpa explained that each set of letters on the list had exactly one letter that referred to a bag that held money. Sherri looked at the list and made some notes. She wrote the letters of the three bags that she thought contained money. Grandpa looked at what she wrote, then gave her the money.

Can you find which three bags held the five-dollar bills? **HINT:** Pick three letters that you think might stand for the three bags with money in them. Then, study the clues. If your choices are correct, exactly one of them will appear in each clue.

AT THE MOVIES

Grandma Ellie's six grandchildren went to the movies on Friday.

- Alex and his sister, Lola
- Eleanor and her sister, Anne
- Martin, an only child
- Tom, an only child

There is one pair of twins in the group. All six grandchildren are different ages, except for the twins. Their ages range from 13 to 17 years old. Read the clues. Determine the ages of the grandchildren and where each child sat in the movie theater.

1. The youngest grandchild is a girl. She is not in the bottom row.
2. The twins are sitting next to each other. They are not the oldest grandchildren.
3. The oldest grandchild is sitting in Section C.
4. Lola and the youngest grandchild are sitting in Section B.
5. Martin is three years older than Anne. They are in the same row.
6. Anne is not the youngest grandchild.
7. Tom is sitting next to the granddaughter who is one year older than he is.
8. The two oldest boys are sitting in the same section.

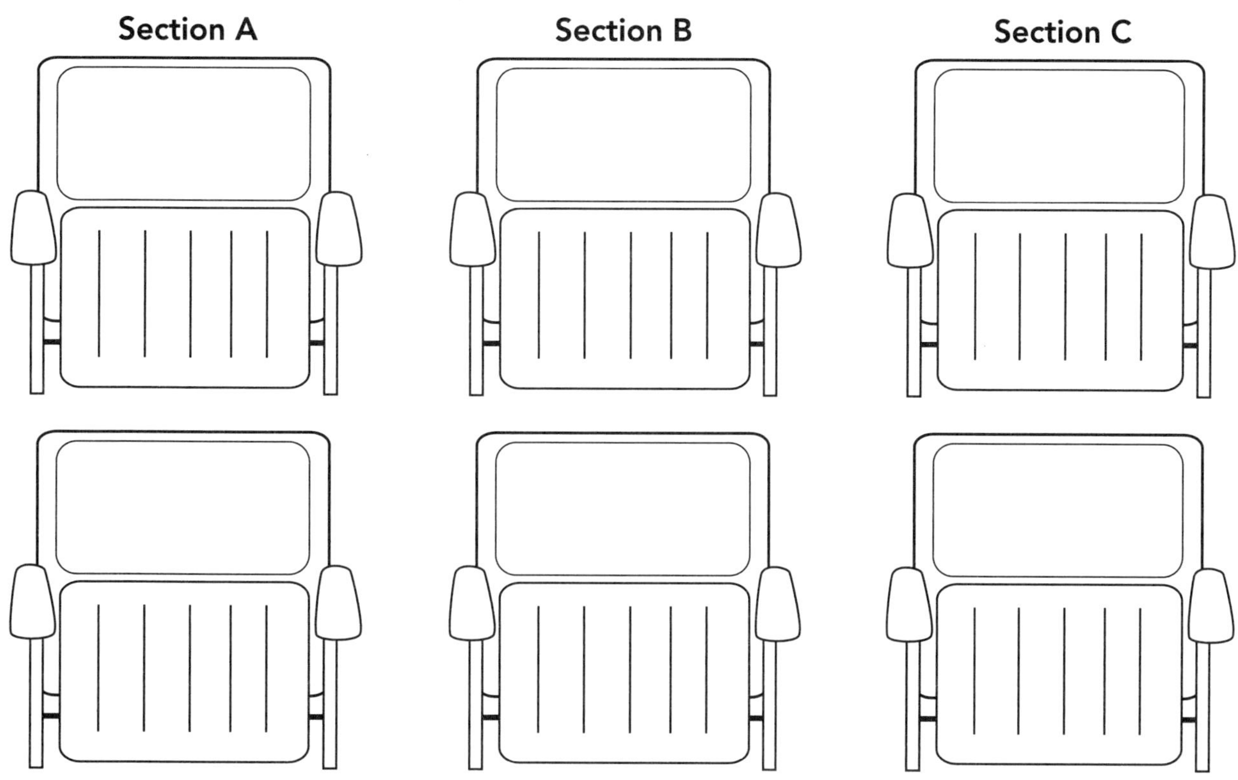

WACKY ADDITION

In the following problems, each letter represents a number between 0 and 9. Determine the value of each letter. HINT: The values of the letters change from problem to problem. For example, the *t* in problem 1 does not have the same value as the *t* in problem 3.

1.
```
   too
 + too
 -----
  blot
```

b = ______ l = ______

o = ______ t = ______

2.
```
   dad
 + dad
 -----
  mama
```

a = ______ d = ______

m = ______

3.
```
   it
 + it
 ----
  tee
```

e = ______ i = ______

t = ______

4.
```
   tot
 + tot
 -----
  yoyo
```

o = ______ t = ______

y = ______

5.
```
   ants
 + ants
 ------
  slant
```

a = ______ l = ______

n = ______ s = ______ t = ______

6.
```
   me
   me
   me
 + me
 ----
   am
```

a = ______ e = ______ m = ______

DIGIT PATTERNS

Study each set of numbers. Circle the number that does not belong. Then, write the common feature that the rest of the numbers share.

1.

263	155
393	(561)
441	177

The product of the two outer digits equals the middle digit.

4.

7485	1559
3627	3524
2446	8271

____

____

2.

246	101
468	718
628	437

____

____

5.

8156	2456
4331	3264
6250	1301

____

____

3.

5735	3824
2612	8216
7963	4137

____

____

6.

5050	2872
3466	6238
4090	1684

____

____

A PATTERN OF ODD NUMBERS

Following the pattern in the first three rows, write odd numbers in the remaining boxes. Then, add the numbers in each row to find an interesting pattern.

							Sum
Row 1:	1						1
Row 2:	3	5					8
Row 3:	7	9	11				______
Row 4:							______
Row 5:							______
Row 6:							______

1. What do you notice about the sums? ______________________________

   ____

2. What would be the sum of the numbers in Row 7? ______________________
3. What would be the sum of the numbers in Row 10? _____________________
4. For which row would the sum be 3,375? ___________________________
5. For which row would the sum be 8,000? ___________________________
6. How would you find the sum of the numbers in any row? ______________

   ____

DIVIDING A SQUARE

A math teacher gave a paper square to each student. She asked her students to draw lines to divide each square into four equal parts.

Most students divided their squares like these examples:

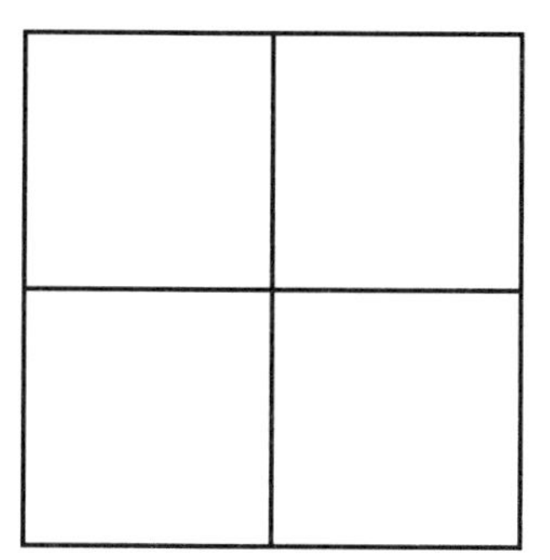

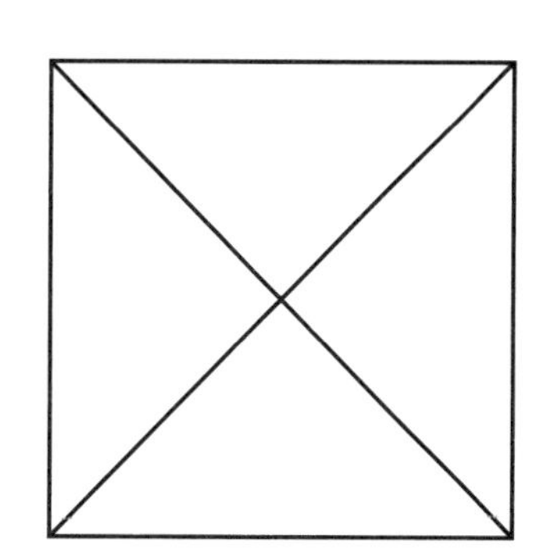

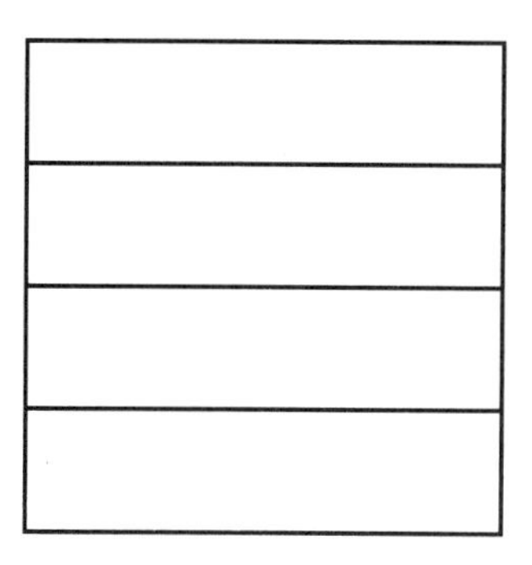

One student folded his square into fourths, then drew lines like this:

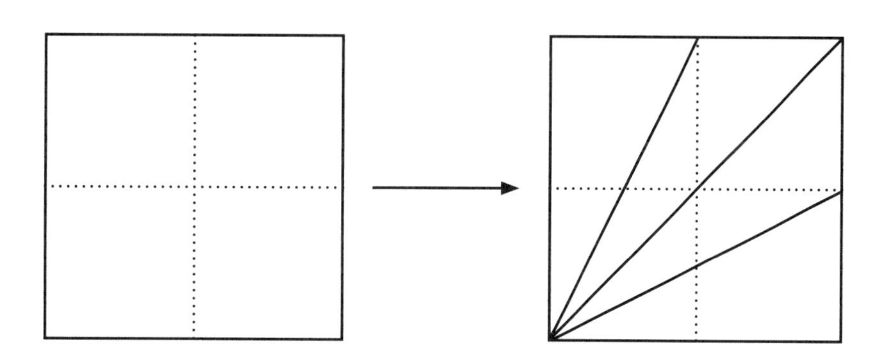

Was the student's paper divided into four equal parts? Explain. ______________________

____

____

____

____

____

CHECKERBOARD SQUARES

What is the total number of squares that can be found on a checkerboard? If a single square's dimensions are 1 x 1, what are the dimensions of the other squares? How many squares of each size are there? Write your answers on the lines. Then, create a formula to solve the problem.

TRIANGLES AND CIRCLES

The circles below are exactly the same size. The dots mark the centers of the circles. Use the circles and the dots to draw two equilateral triangles. HINT: The sides of one triangle will be twice as long as the sides of the other triangle.

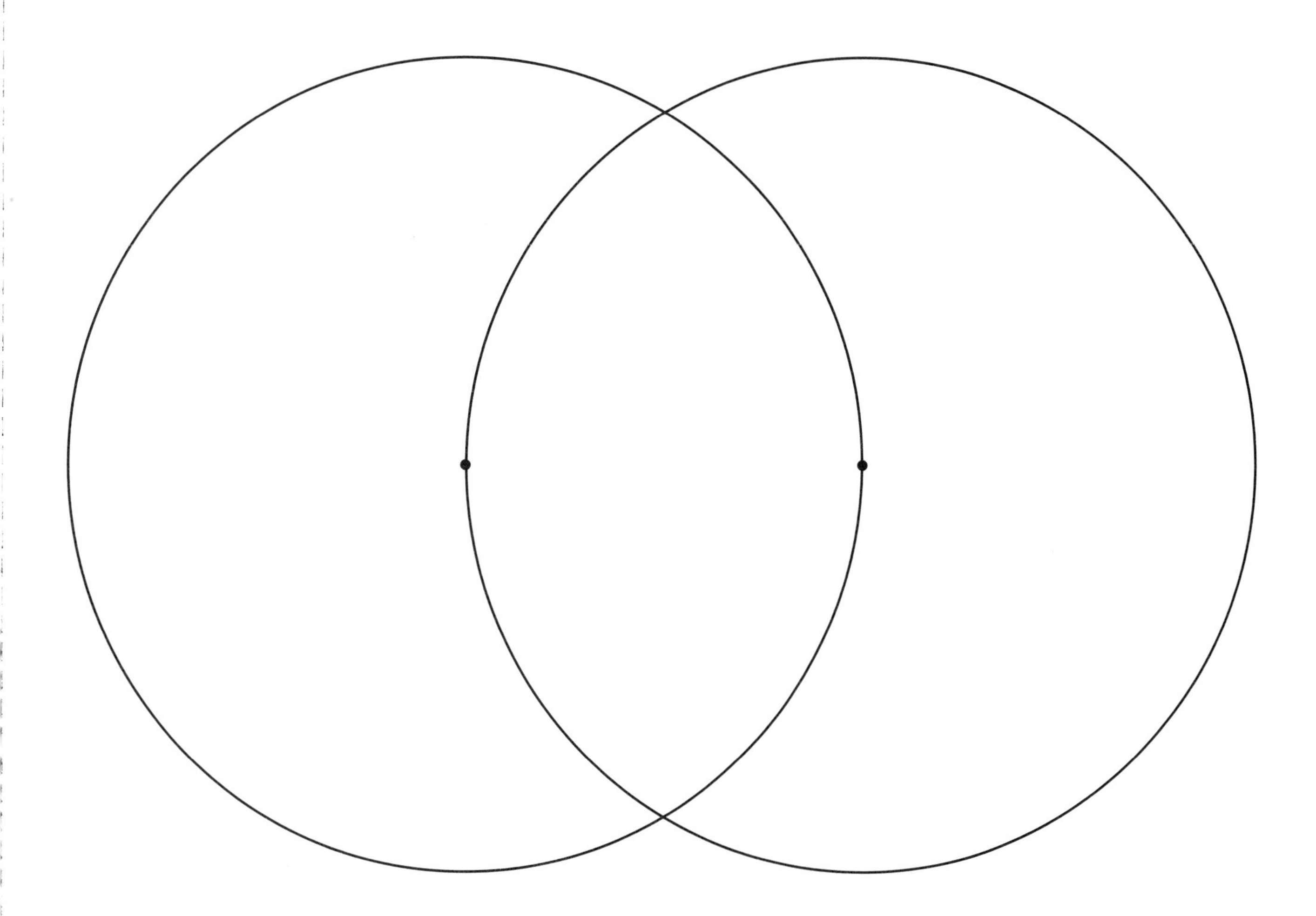

PAINTERS AT WORK

Dan painted the outside of this cube. The faces of the smaller cubes that touch other cubes are not painted. Think about what the cube will look like when it is taken apart. Then, answer each question.

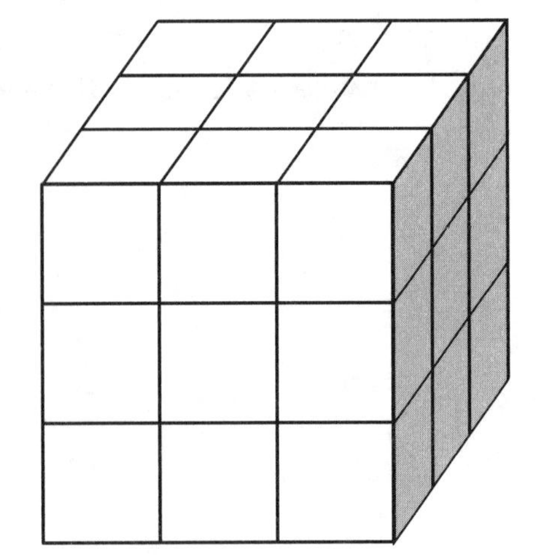

1. How many cubes will have 0 faces painted? ________
2. How many cubes will have 1 face painted? ________
3. How many cubes will have 2 faces painted? ________
4. How many cubes will have 3 faces painted? ________

Fran painted the outside of this cube. The faces of the smaller cubes that touch other cubes are not painted. Think about what the cube will look like when it is taken apart. Then, answer each question.

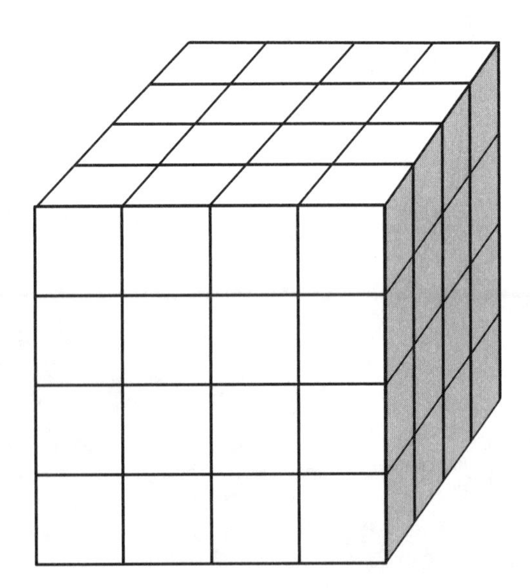

5. How many cubes will have 0 faces painted? ________
6. How many cubes will have 1 face painted? ________
7. How many cubes will have 2 faces painted? ________
8. How many cubes will have 3 faces painted? ________

TRIANGLE HUNT

Lynn, Jeremy, and Megan made different designs with tiles.

1. How many triangles can be found in Lynn's design? __________

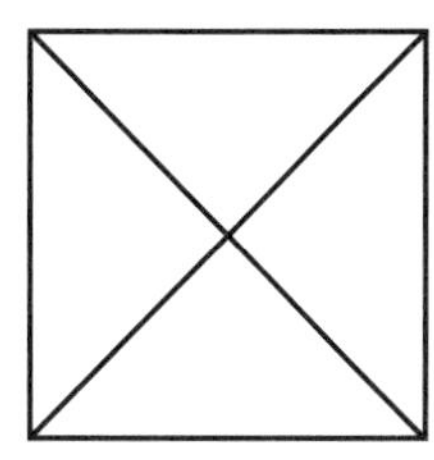

2. How many triangles can be found in Jeremy's design? __________

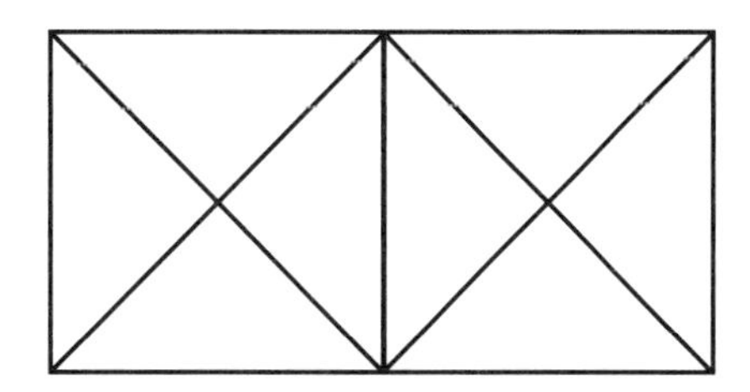

3. How many triangles can be found in Megan's design? __________

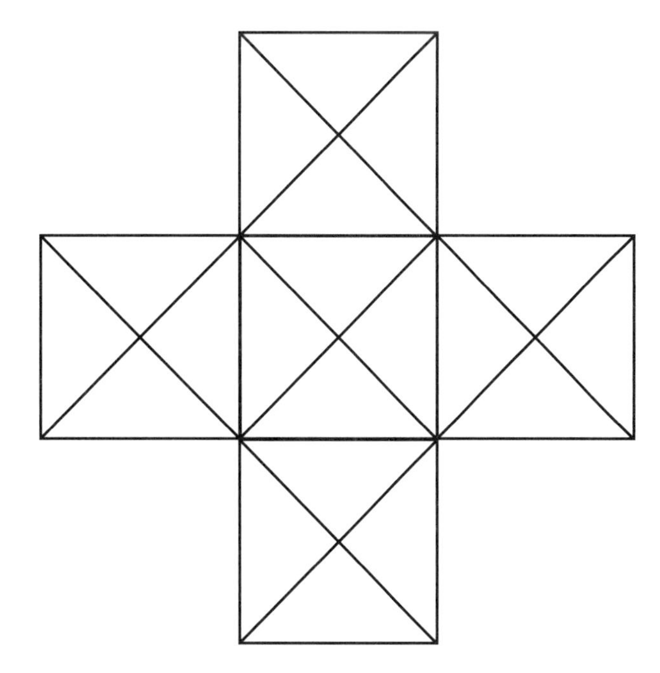

CONNECT THE DOTS

Connect all of the dots by drawing four straight lines. You may start from any place on the paper. Once you begin drawing, do not lift your pencil off of the paper until you are finished. Lines can cross, but do not go back over any lines that have already been drawn.

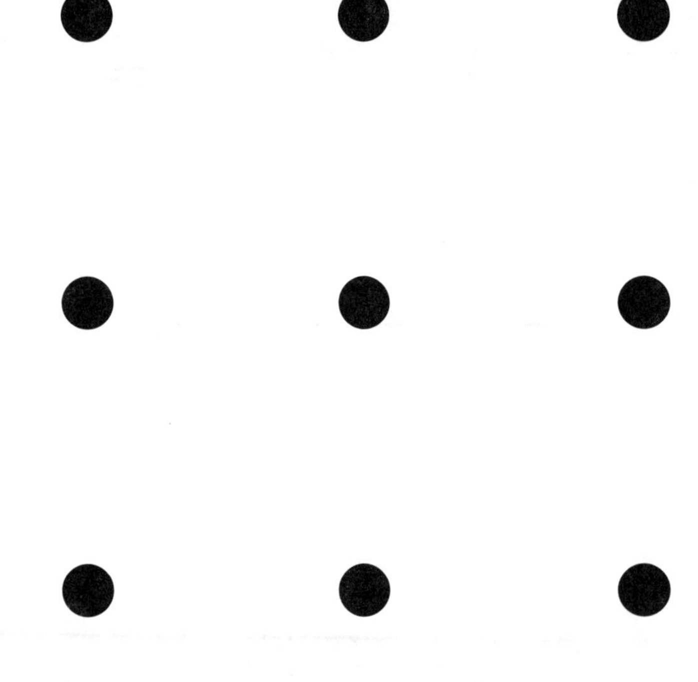

THE TOTAL

Use the clues and the table to find the fraction that each person illustrated.

1. Neither Nathan nor Gabe illustrated the largest fraction.
2. Jayla doubled $\frac{1}{8}$ to get her fraction.
3. Paige's fraction is less than Gabe's fraction.
4. Maddie illustrated the smallest fraction.
5. José subtracted $\frac{1}{4}$ from $\frac{5}{6}$ to get his fraction.
6. José subtracted Jayla's fraction from Wendy's fraction to get his fraction.
7. The sum of José's, Wendy's, Nathan's, and Jayla's fractions is 2.
8. The difference between Nathan's and George's fractions is $\frac{1}{9}$.

	$\frac{1}{5}$	$\frac{7}{12}$	$\frac{3}{8}$	$\frac{5}{6}$	$\frac{1}{3}$	$\frac{5}{10}$	$\frac{1}{4}$	$\frac{4}{9}$
Gabe								
George								
Jayla								
José								
Maddie								
Nathan								
Paige								
Wendy								

Jayla's fraction: ____________

Gabe's fraction: ____________

George's fraction: ____________

José's fraction: ____________

Maddie's fraction: ____________

Nathan's fraction: ____________

Paige's fraction: ____________

Wendy's fraction: ____________

Use the clues and the table to match the band members with their instruments. HINT: To solve this puzzle, you may need to research musical instruments on the Internet, in an encyclopedia, or at your school or local library.

1. Chad, Jake, and Valerie play woodwinds.
2. Abby, Ian, and Olivia do not play brass instruments.
3. Chad does not play a saxophone.
4. Olivia memorized the tune for her percussion instrument.
5. Jake's instrument does not have a reed.
6. Abby does not play the bass drum.
7. Ian and Joseph play instruments in different families, but both instruments have the lowest range of pitches in their families.
8. Sarah plays a brass instrument but does not like the long slide on Zora's horn.

	bass drum	bells	clarinet	flute	saxophone	snare drum	trombone	trumpet	tuba
Abby									
Chad									
Ian									
Jake									
Joseph									
Olivia									
Sarah									
Valerie									
Zora									

Abby plays the ______________________ .

Sarah plays the ______________________ .

Chad plays the ______________________ .

Ian plays the ______________________ .

Jake plays the ______________________ .

Joseph plays the ______________________ .

Olivia plays the ______________________ .

Valerie plays the ______________________ .

Zora plays the ______________________ .

COULD BE THE ANSWER

Could the given answer be correct? Use the order of operations. Add parentheses and brackets as needed to make each equation true.

A. $4 + 5 \times 8 - 2 \times 4 = 64$

B. $7 + 3 - 2 \times 3 + 2 - 2 \times 8 = 24$

C. $15 - 3 + 7 \times 5 - 2 + 2 = 35$

D. $8 \div 5 - 1 + 2 + 5 \times 9 - 3 + 4 = 14$

The parentheses in the second part of each equation are given. Add parentheses and brackets as needed in the first part of each equation to make it true.

E. $8 + 2 \times 3 + 6 \div 9 = 14 \div 7 + 12 \div (4 + 2)$

F. $8 + 2 + 5 \div 5 + 3 + 12 \div 3 = 14 \div (2 + 5) + 24 \div (1 + 5)$

G. $9 - 3 + 8 - 2 \times 3 = 3 \times 3 + (11 - 6 \times 2)$

H. $7 - 1 \times 3 - 10 \div 4 + 9 - 8 = 3 \times (11 - 7) + (6 - 4) \times 2$

Place function symbols, parentheses, and brackets in these equations to make them true.

I. $6 \quad 1 \quad 3 \quad 3 \quad 6 = 4$

J. $9 \quad 3 \quad 5 \quad 4 \quad 1 \quad 8 = 3$

K. $4 \quad 5 \quad 2 \quad 5 \quad 12 \quad 8 = 2$

L. $9 \quad 6 \quad 5 \quad 8 = 27$

M. $\frac{1}{2} \quad \frac{1}{4} \quad \frac{2}{3} \quad \frac{1}{4} \quad \frac{6}{8} = 1$

N. $2.4 \quad \frac{1}{2} \quad \frac{3}{6} \quad 6.6 \quad \frac{1}{6} = 0.3$

O. $5.02 \quad 3 \quad \frac{2}{6} \quad 0.4 \quad 3 = 2.82$

P. $5\frac{3}{4} \quad 1.47 \quad 0.04 \quad 4\frac{5}{6} \quad 2.4 \quad 0.6 = 216.6$

COULD BE THE ANSWER

How many different answers can you find to these problems by adding parentheses in different places? Write six options for each equation.

1. $6 + 4 \times 2 \div 5 - 3 \times 16 \div 2 + 6 = ?$

A. $6 + 4 \times 2 \div 5 - 3 \times 16 \div 2 + 6 =$ ______

B. $6 + 4 \times 2 \div 5 - 3 \times 16 \div 2 + 6 =$ ______

C. $6 + 4 \times 2 \div 5 - 3 \times 16 \div 2 + 6 =$ ______

D. $6 + 4 \times 2 \div 5 - 3 \times 16 \div 2 + 6 =$ ______

E. $6 + 4 \times 2 \div 5 - 3 \times 16 \div 2 + 6 =$ ______

F. $6 + 4 \times 2 \div 5 - 3 \times 16 \div 2 + 6 =$ ______

2. $9 - 3 \times 2 + 1 + 2 \times 12 \div 6 - 2 = ?$

G. $9 - 3 \times 2 + 1 + 2 \times 12 \div 6 - 2 =$ ______

H. $9 - 3 \times 2 + 1 + 2 \times 12 \div 6 - 2 =$ ______

I. $9 - 3 \times 2 + 1 + 2 \times 12 \div 6 - 2 =$ ______

J. $9 - 3 \times 2 + 1 + 2 \times 12 \div 6 - 2 =$ ______

K. $9 - 3 \times 2 + 1 + 2 \times 12 \div 6 - 2 =$ ______

L. $9 - 3 \times 2 + 1 + 2 \times 12 \div 6 - 2 =$ ______

3. $24 \div 8 - 2 + 2 \times 12 - 7 + 1 \times 5 = ?$

M. $24 \div 8 - 2 + 2 \times 12 - 7 + 1 \times 5 =$ ______

N. $24 \div 8 - 2 + 2 \times 12 - 7 + 1 \times 5 =$ ______

O. $24 \div 8 - 2 + 2 \times 12 - 7 + 1 \times 5 =$ ______

P. $24 \div 8 - 2 + 2 \times 12 - 7 + 1 \times 5 =$ ______

Q. $24 \div 8 - 2 + 2 \times 12 - 7 + 1 \times 5 =$ ______

R. $24 \div 8 - 2 + 2 \times 12 - 7 + 1 \times 5 =$ ______

FARMER'S MARKET

Use the clues to decide which fruit or vegetable goes in each basket. Write the name of the correct food in each basket. HINT: One basket is empty.

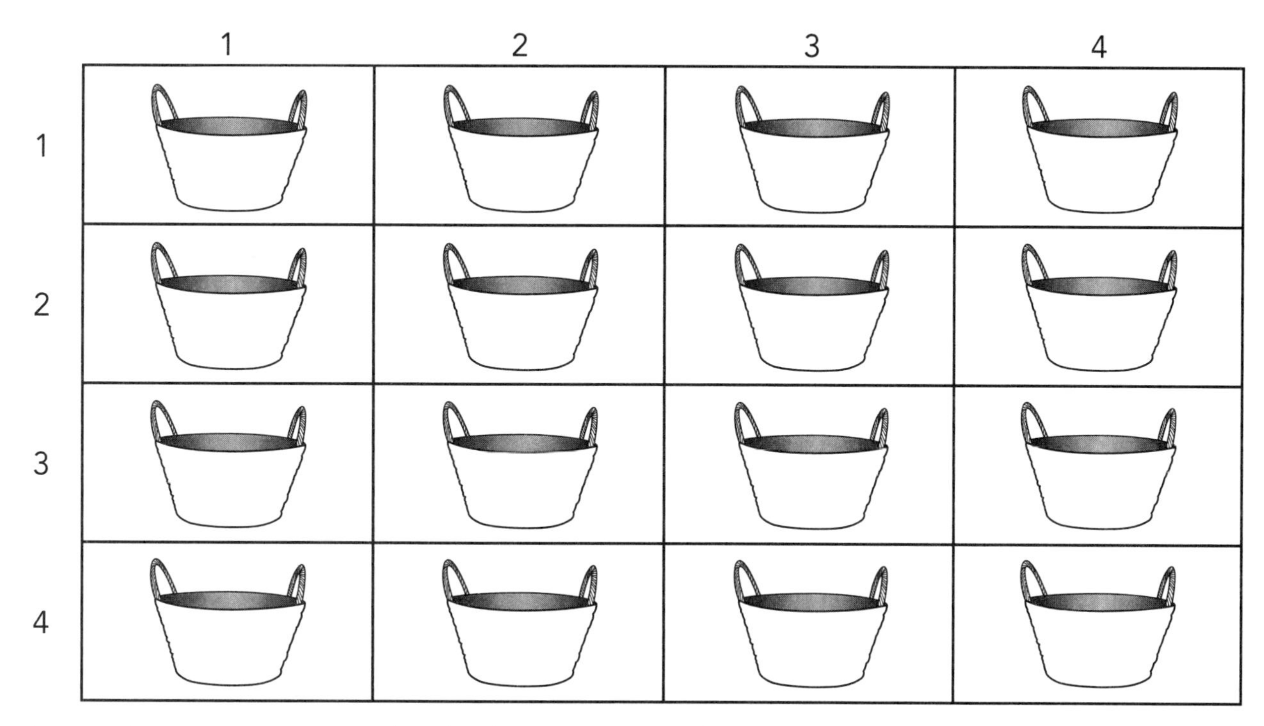

1. The apple is in the bottom left basket.
2. The broccoli and the banana are in the same row.
3. The pear is not is the same row or column as the apple.
4. No vegetables are in the third column.
5. The banana is in the same row as the pear and the same column as the apple.
6. Place the orange in a basket directly between the banana and the pear.
7. The apricot is in one of the first two columns.
8. The carrot is in the same column as the broccoli.
9. In the pear's column is the basket with the grapes.
10. No fruits are in the first row.
11. The grapes are in the same row as the apple and the carrot.
12. The beans are in the column before the pear.
13. Corn and celery share the same row as the beans.
14. The strawberries are not in the first two rows.
15. The celery is in a column after the corn.
16. In the grapes' row is the lemon.
17. The lemon and the orange share a column with the cauliflower.
18. The orange is not in the third row.
19. Corn is in the same column as the apple. It is in the row before the pear.
20. The plum is in the column before the carrot. It is in the same row as the cauliflower.

PUZZLE FILL-IN

Fill in the puzzle with four-digit products from multiplication problems. Each problem's factors must be two-digit numbers. Write each problem on the correct clue line.

Clues:

Down

3. ____________________
4. ____________________
5. ____________________
7. ____________________
8. ____________________
9. ____________________
10. ____________________
12. ____________________
13. ____________________

Across

1. ____________________
2. ____________________
4. ____________________
5. ____________________
6. ____________________
7. ____________________
9. ____________________
10. ____________________
11. ____________________
12. ____________________

STEPS

Add each pair of adjacent numbers and write the sum in the box above them. Use the given numbers to find the missing numbers. Continue until each set of steps is completed. Reduce each fraction to its simplest form. Change improper fractions to mixed numbers.

1.

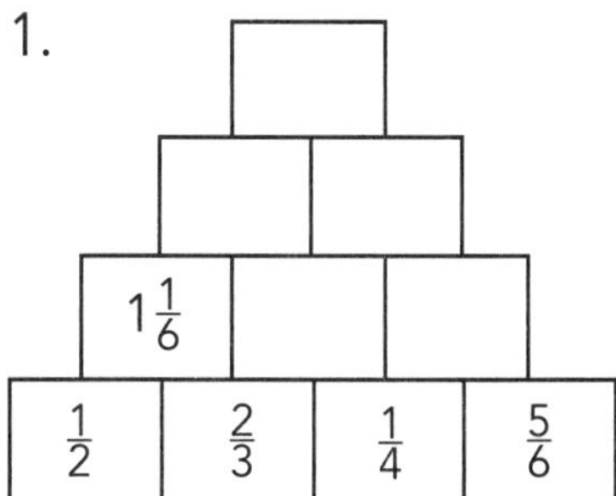

2.

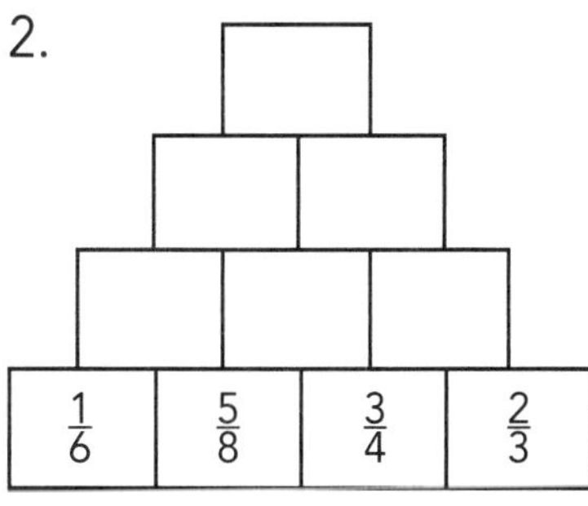

3.

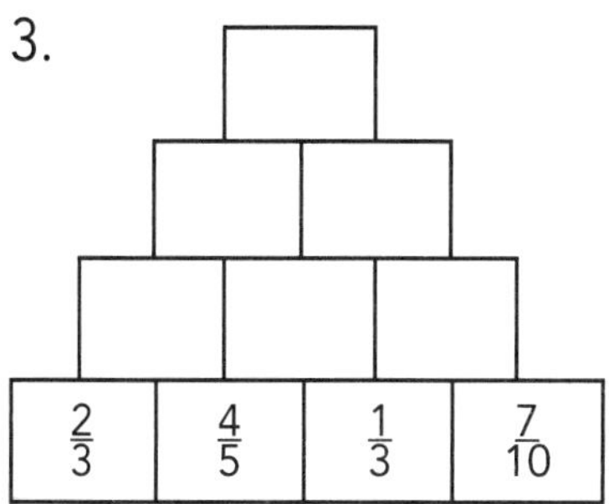

4.

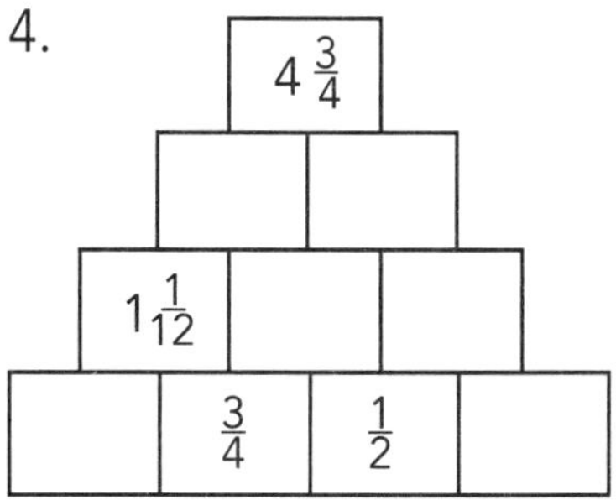

5.

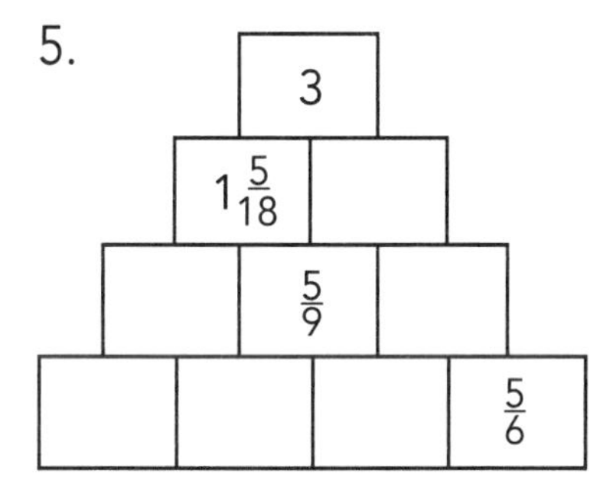

6.

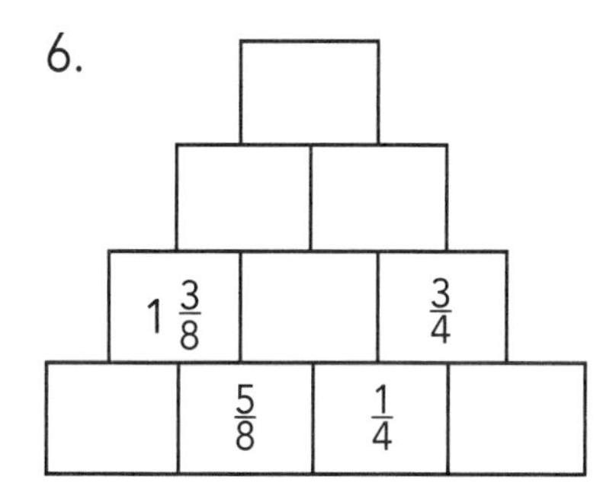

7.

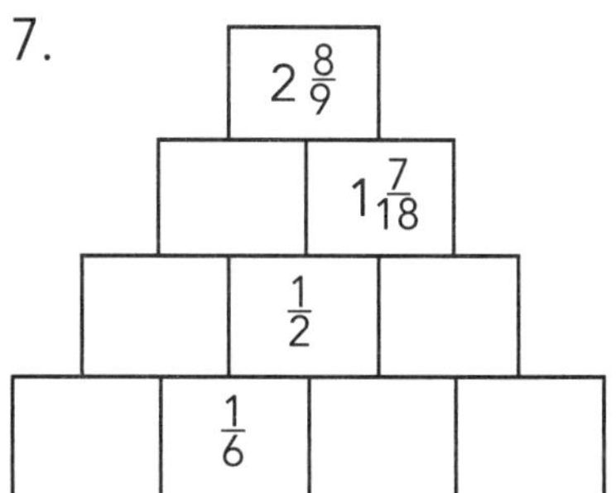

8.

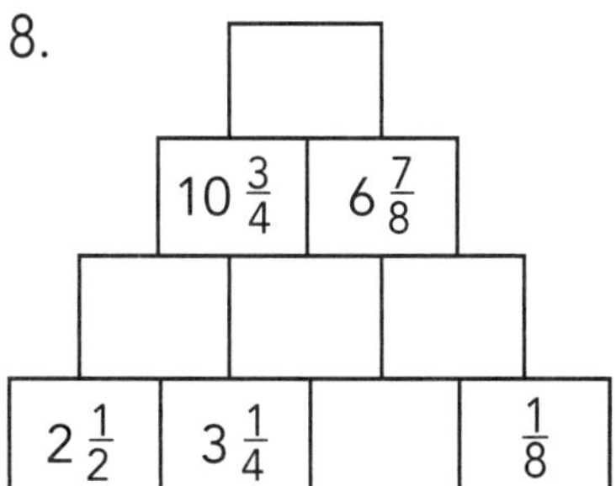

9.

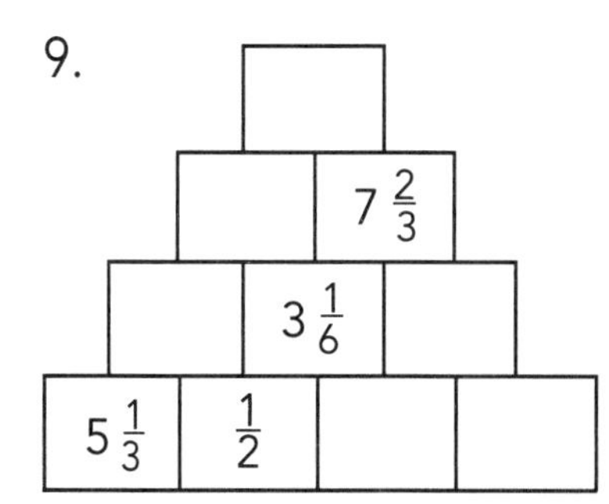

MISSING STEPS

Add your way to the top. Add each pair of adjacent numbers and write the sum in the box above them. Use each of the numbers given once to fill in the first row of steps.

1, 2, 3, 4, 5, 6, 7, 8, 9

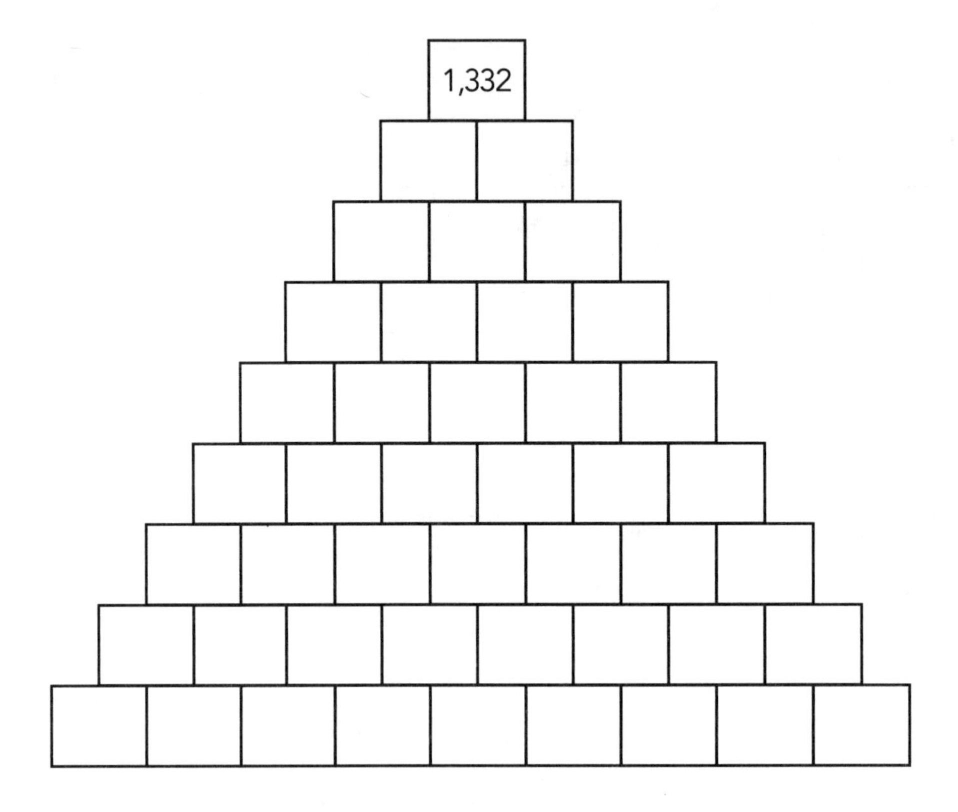

NUMBER RIDDLES

Solve each number riddle. Write the answer on the line.

1. Subtract 16 from me. Subtract 6, then divide by 4. You get -5.
   What number am I? ______________________

2. Divide me by 2. Subtract 1, then add 10. You get 6.
   What number am I? ______________________

3. Divide 24 by me. Then, subtract 5 and multiply by 4. You get -12.
   What number am I? ______________________

4. Divide me by 8, add 10, then multiply by 3. Subtract 11 to get 7.
   What number am I? ______________________

5. Add 2 to me, multiply by 3, then subtract 3. Divide by 8, then subtract 7. You get -4.
   What number am I? ______________________

6. Subtract 9 before adding 5 to me, then double the answer. Add 2, then divide by 7 and subtract 2. You get 0.
   What number am I? ______________________

7. Add 2, then add 4 more. Subtract 6. Add 12, then divide by 3. You get -3.
   What number am I? ______________________

8. Subtract 1 from me. Add 9, then subtract 2. Add 9 more. Multiply by 9, then divide by 6. You get 3.
   What number am I? ______________________

9. Subtract 6 from me, then add 4 and multiply by 2. Add 10 and divide by 2. You get 4.
   What number am I? ______________________

10. Divide me by 5, then add 12. Divide by 3 and multiply by 6. You get 18.
    What number am I? ______________________

11. Subtract 7 from me, then multiply me by 5 and add 8. Multiply by 3 and you get -6.
    What number am I? ______________________

12. Add 12 to me, multiply by 4, and subtract 6. Divide by 2, and you get 7.
    What number am I? ______________________

MONEY MIX-UP

Use the clues and the table to determine who has each set of money. Monetary units used include: pennies, nickels, dimes, quarters, 50-cent pieces, and 1-dollar bills.

1. Jamison can make his total using only quarters.
2. Hailey has more money than Melanie.
3. Trina said, "I have at least 1 penny, but fewer than 3 pennies."
4. Samuel has more than Derek or Trina.
5. Katie has less money than Ruth.
6. Samuel has 2 bills and 7 coins.
7. Melanie's money includes 8 fifty-cent pieces.
8. If Katie's total is subtracted from Jamison's total, the result is close to Ruth's total.
9. Derek does not have the smallest amount.
10. Ruth's total is greater than Ian's total but less than Derek's total.

	$0.89	$1.25	$1.77	$2.61	$2.88	$3.00	$3.53	$4.34	$4.50
Derek									
Hailey									
Ian									
Jamison									
Katie									
Melanie									
Ruth									
Samuel									
Trina									

Derek has ______________________________ .

Hailey has ______________________________ .

Ian has ______________________________ .

Jamison has ______________________________ .

Katie has ______________________________ .

Melanie has ______________________________ .

Ruth has ______________________________ .

Samuel has ______________________________ .

Trina has ______________________________ .

WORD FRAGMENTS

The following letter combinations appear in the answer to each clue. You may add letters to the beginnings of the fragments, the ends, or both. You may not add letters within the fragments themselves.

1. laus – believable, reasonable ____________
2. bdic – to give up a position ____________
3. rres – relating to land ____________
4. cari – experienced through another ____________
5. lemi – an imperfection, flaw ____________
6. olog – arranged in order of time ____________
7. tton – overindulgence, as in food and drink ____________
8. minat – what something progresses toward ____________
9. elib – intentional ____________
10. tran – not necessary ____________
11. rarc – a system with ranked groups ____________
12. ndig – originating in a region ____________
13. udic – sound or sensible ____________
14. ealo – filled with eagerness for something ____________
15. thar – in a state of sluggishness or apathy ____________
16. morph – a change of form, shape, or substance ____________
17. adic – wandering from place to place ____________
18. sol – no longer used, out of date ____________
19. tho – conventional ____________
20. and – a state of perplexity ____________
21. sere – to find good things without looking for them ____________
22. miss – easily yielding to authority ____________
23. top – an imaginary and remote place of perfection ____________
24. unct – connection ____________
25. him – fanciful ____________
26. tuit – happening by chance ____________

WORD FRAGMENTS

The following letter combinations appear in the answer to each clue. You may add letters to the beginnings of the fragments, the ends, or both. You may not add letters within the fragments themselves.

1. ffl – rich; aristocratic ____________________
2. ccl – praise ____________________
3. dgm – opinion, usually one with greater authority ____________________
4. viat – to make easier or to lessen, as in pain ____________________
5. rpho – having no definite, solid structure ____________________
6. ajo – to convince through gentle persistence ____________________
7. hag – a feeling of shame or embarrassment ____________________
8. xte – the ability to use one's body with great skill and ease ____________________
9. aes – possessing beauty or style ____________________
10. xac – to make worse, as in pain or a situation ____________________
11. sco – a large failure; debacle ____________________
12. rru – loving to talk, especially about trivial matters ____________________
13. rbin – something that announces or foreshadows something else ____________________
14. ybr – a mixture of two things with different origins ____________________
15. mpec – without fault ____________________
16. sci – easily angered; hot-tempered ____________________
17. nth – a large maze ____________________
18. ckkn – a cheap, unimportant item ____________________
19. aus – large, above-ground tomb ____________________
20. imat – to purify something ____________________
21. nde – chaos, disorder, confusion ____________________
22. uper – an extremely poor person ____________________
23. enul – the second-to-last item in a series ____________________
24. don – an assumed name, especially as an author ____________________
25. ueu – a waiting line ____________________
26. mphl – small booklet, often used for advertisements ____________________

WORD FRAGMENTS

The following letter combinations appear in the answer to each clue. You may add letters to the beginnings of the fragments, the ends, or both. You may not add letters within the fragments themselves.

1. noun – to criticize publicly ______
2. cent – to stress, highlight ______
3. hem – something of tremendous power or size ______
4. aphy – the arrangement of dances ______
5. nart – incapable of expressing oneself clearly through speech ______
6. hear – feeling a loss of spirit ______
7. bell – to add details to, enhance ______
8. berg – astounded ______
9. rug – thrifty, economical ______
10. cat – profession ______
11. row – distressing, vexing ______
12. the – possible, but unproven ______
13. rid – to cut down, shorten ______
14. lant – extremely joyful, happy ______
15. udo – praise for an achievement ______
16. min – brightly shining ______
17. ticu – extremely careful with details ______
18. min – foreshadowing evil ______
19. ram – greatest in importance, rank, or character ______
20. put – of good reputation ______
21. ran – calm ______
22. her – logically consistent, intelligible ______
23. uixo – idealistic, impractical ______
24. surr – kept secret ______
25. ferv – fizzy ______
26. etho – an excessive amount ______

COMMON LINKS

The words in each set share a common associated word. This common word may come before or after each listed word. Determine the common word for each set.

1. rabbit, black, hammer: ______________________
2. dinner, round, end: ______________________
3. ice, axe, lock: ______________________
4. detector, screen, stack: ______________________
5. front, salt, bed: ______________________
6. baseball, size, knee: ______________________
7. over, tuning, salad: ______________________
8. date, ball, one-act: ______________________
9. mate, list, bounced: ______________________
10. worthy, whole, book: ______________________
11. game, down, case: ______________________
12. dream, break, light: ______________________
13. fall, mare, good: ______________________
14. full, harvest, blue: ______________________
15. cellar, window, rain: ______________________

RHYME TIME

Each of the following two-word phrases rhymes with a common noun. Use the category clues to decipher the nouns hidden by the rhymes. The number in parentheses following each clue tells how many words make up the answer.

Around the House

1. charm fare (1) ____________________
2. toffee gable (2) ____________________
3. wrath vowel (2) ____________________
4. shows fire (2) ____________________
5. peeling flan (2) ____________________
6. hurtin' Todd (2) ____________________
7. freebie net (2) ____________________
8. pilfer stare (1) ____________________
9. shawl caper (1) ____________________
10. sanity nearer (2) ____________________
11. math cub (1) ____________________
12. pear mace (1) ____________________

At the Office

13. sacks unclean (2) ____________________
14. daughter jeweler (2) ____________________
15. toffee faker (2) ____________________
16. gorilla boulder (2) ____________________
17. slack cuisine (2) ____________________
18. reading groom (2) ____________________
19. scorner doffmiss (2) ____________________
20. razor splinter (2) ____________________

Each of the following problems is a common phrase or saying. Each word in the phrase has had one letter changed so that a new word is created. Decipher each phrase and write it on the line.

1. Taste lakes paste ______
2. I foot ant him Monet arm loon ported ______
3. Ale roans leak do home ______
4. Brood it thicket that hater ______
5. Won't coy oven shilled silk ______
6. Bit at I riddle ______
7. Hit bard as horse thaw him bile ______
8. Whey in mains, at sours ______
9. Take way white she pun shires ______
10. An fob I peony, is nor I wound ______
11. Fever fudge I boom be if's rover ______
12. To mat in on inland ______
13. To pail, so gait ______
14. I roiling shone fathers to most ______
15. Home way nit guilt on I dam ______

WORD EQUATIONS

The puzzles below are word equations. In each puzzle, the bold letters represent specific words. When combined, the letters and numbers represent common facts or phrases. Write each phrase on the line.

1. 12 **M** in a **Y** ______
2. 4 **Q** in a **D** ______
3. 9 **J** on the **U.S. S C** ______
4. 29 **D** in **F** in a **L Y** ______
5. 24 **K** in **P G** ______
6. 535 **L** in the **U.S. C** ______
7. 23 **C** in **N A** ______
8. 5 **T** on a **H F** ______
9. 90 **D** in a **R A** ______
10. 360 **D** in a **C** ______
11. 7 **W** of the **W** ______
12. 18 **H** on a **G C** ______
13. 3 **W** on a **T** ______
14. 52 **C** in a **D**, not counting the **J** ______
15. 1760 **Y** in a **M** ______
16. 8 **P** in the **S S** ______
17. 6 **S** on a **D** ______
18. 100 **D** in a **M** ______
19. 3 **C** on a **S** ______
20. 5 **S** on a **P** ______
21. 100 **C** in a **M** ______
22. 7 **C** on **E** ______

ANALOGIES

Circle the letter in front of the answer that correctly completes each analogy.

1. Desert : rainforest :: ______ : ravine

   A. ocean B. canyon C. plateau D. mountain

2. tasteless : bland :: auspicious : ______

   A. foreboding B. favorable C. trepidation D. suspicious

3. sight : eye :: ____ : fingers

   A. play B. touch C. feel D. move

4. bird : nest :: rabbit : ______

   A. field B. den C. carrot D. burrow

5. mobile phone : battery :: human : _____

   A. food B. clothing D. shelter D. brain

6. poet : verses :: ____________________

   A. cooper : shoes B. cobbler : hats
   C. novelist : music D. cartographer : maps

7. valiant : courage :: ____________________

   A. chipper : melancholy B. wrathful : boredom
   C. tyrannical : power D. frightened : effrontery

8. gale : wind :: ____________________

   A. deluge : rain B. snow : blizzard
   C. flood : tidal wave D. frostbite : cold

9. virtuoso : mediocre :: ____________________

   A. novice : inexperienced B. talented : gifted
   C. recluse : sociable D. nomad: itinerant

10. Dexterity : nimble :: ____________________

    A. integrity : duplicitous B. complacent : eager
    C. novel : pamphlet D. hubris: arrogance

BOOK BENDER

A bookshelf holds eight books. The cover of each book is a different color. Use the clues and the table to find the order of the eight books on the shelf.

1. The red book is directly to the left of the white book.
2. The orange book is either the first book or the last book.
3. The teal book and the brown book are next to each other.
4. The fifth book is either green or purple.
5. The green book is directly between the orange book and the teal book.
6. The teal book's order number is a number that, when spelled out, begins with the same letter.
7. The white book is either the first or last book.

	1	2	3	4	5	6	7	8
Red								
Brown								
Teal								
Yellow								
Orange								
Purple								
Green								
White								

CAR CONUNDRUM

Seven friends—Andy, Brenda, Carl, Debbie, Emily, Frank, and Gerald—decide to go to a baseball game. However, because different people want to leave at different times, the friends decide to take three cars. Use the clues and the table to determine who rode together to the baseball game.

1. No car has fewer than two people in it.
2. Carl will not ride in a car that has more than two people in it, himself included.
3. Emily and Brenda are neighbors, and they always ride together.
4. Gerald will only ride in a car with at least two other people in it.
5. Andy and Brenda are both driving their own cars.
6. Frank will only ride in a car if Debbie is also in it.
7. No car has only two girls riding in it.

	Car A	Car B	Car C
Andy			
Brenda			
Carl			
Debbie			
Emily			
Frank			
Gerald			

TWISTED WORDS

Find the seven-letter word hidden in each grid. The letters in the word must touch horizontally, vertically, backward, or forward, but not diagonally. Write each word on the line provided.

1.

o	e	l	t
g	r	y	i
t	a	s	r
u	m	m	a

____

2.

l	o	l	t
g	t	i	i
t	r	p	a
e	m	e	c

____

3.

s	e	u	t
e	p	i	s
d	a	c	t
u	s	m	a

____

4.

b	e	h	w
g	t	i	r
t	i	n	r
e	m	g	a

____

THINK LINKS

Write a pair of rhyming words that means the same thing as each clue.

1. displeased supervisor ______________________________
2. angry boy ______________________________
3. eager adolescent ______________________________
4. pale twosome ______________________________
5. genuine transaction ______________________________
6. chocolate-colored dress ______________________________
7. amusing rabbit ______________________________
8. chief hurt ______________________________
9. benevolent intellect ______________________________
10. uninteresting seabird ______________________________

The following clues are alliterations. In each blank, write a pair of words that means the same, or about the same, thing as the clue. The answer must be an alliteration.

11. big boy ______________________________
12. forlorn father ______________________________
13. smelly socks ______________________________
14. grouchy grandfather ______________________________
15. peppy pooch ______________________________
16. filthy finger ______________________________
17. tragic tune ______________________________
18. misplaced money ______________________________
19. short sailor ______________________________
20. positive pal ______________________________

BEASTLY WORDS

Find an animal word in each word below. The word may be found by rearranging letters. HINT: Not all of the letters in each word must be found in the new word.

1. gnaws ______
2. winter ______
3. kelp ______
4. autumn ______
5. toward ______
6. noodle ______
7. laser ______
8. nodding ______
9. jungle ______
10. omelet ______
11. dribble ______
12. snacker ______
13. shore ______
14. toga ______
15. harks ______
16. reed ______
17. trowel ______
18. flower ______
19. spring ______
20. forget ______

THAT'S "PUN-NY"

Write a pun to answer each question.

1. What bird is a grumpy individual? ______________________
2. What plant is a superb feline? ______________________
3. What flower is best for playing the trumpet? ______________________
4. What insect is a lemon-colored garment? ______________________
5 What bird is a nocturnal windstorm? ______________________
6. Which insect is someone who gives medicine to felines? ______________________
7. What plant has something to do with clocks? ______________________
8. What mammal is part of a chain? ______________________
9. What bird means *very fast*? ______________________
10. What insect is a listening device attached to a piece of furniture? ______________________
11. What flower is a country of automobiles? ______________________
12. What tree is not me? ______________________
13. What bird is a baseball outfielder? ______________________
14. What plant is the bite of a feline? ______________________
15. What flower says hello to Cynthia? ______________________

HIDDEN WORDS

Circle the name of a body part hidden in each sentence.

1. The monkey eats bananas.
2. The advertisement is in today's newspaper.
3. Frank needs a haircut.
4. Purple glasses are cool!
5. The car made a turn.
6. Hannah and Conley are friends.
7. Sue's kindness did not go unnoticed.
8. Which ink is best for drawing?

Circle the number word hidden in each sentence.

9. The mouse ventured into the open.
10. He ate nuts and berries.
11. The ice froze Rose's toes.
12. We will go next week.
13. Dad left work already.
14. The clarinet came with reeds.
15. Mrs. Smith left off our names.
16. If I've got time, I can help you.

CRYPTOGRAM

Use the code to write the letters in the puzzle. The completed puzzle will reveal a quotation by Abraham Lincoln.

A	B	C	D	E	G	H	I	L	M	N	O	P	R	S	T	U	Y	W
7	9		18		21	14	3				11	6		16		10		

__ __ __ __ __ __   __ __ __ __   __ __
7 26 5 7 20 16   9 12 7 2   3 8

__ __ __ __   __ __ __ __   __ __ __ __
15 3 8 18   23 14 7 23   20 11 10 2

__ __ __   __ __ __ __ __ __ __ __ __ __
11 5 8   2 12 16 11 26 10 23 3 11 8

__ __   __ __ __ __ __ __ __   __ __
23 11   16 10 1 1 12 12 18   3 16

__ __ __ __   __ __ __ __ __ __ __ __ __
15 11 2 12   3 15 6 11 2 23 7 8 23

__ __ __ __   __ __ __   __ __ __
23 14 7 8   7 8 20   11 8 12

__ __ __ __ __ .
23 14 3 8 21

CRYPTOGRAM

Use the code to write the letters in the puzzle. The completed puzzle will reveal a quotation by Albert Einstein.

A	B	C	D	E	F	G	H	I	K	L	M	N	O	P	R	S	T	W
		22	3				6		17	12		7	16	9		4		

__ __ __ __ __ __ __ __ __ __ __   __ __
20 15 19 5 20 7 19 2 20 16 7   20 4

__ __ __ __   __ __ __ __ __ __ __ __ __
15 16 14 25   20 15 9 16 14 2 19 7 2

__ __ __ __   __ __ __ __ __ __ __ __ __.
2 6 19 7   17 7 16 11 12 25 3 5 25

__ __ __   __ __ __ __ __ __ __ __ __
18 16 14   17 7 16 11 12 25 3 5 25

__ __   __ __ __ __ __ __ __,
20 4   12 20 15 20 2 25 3

__ __ __ __ __ __ __
11 6 25 14 25 19 4

__ __ __ __ __ __ __ __ __ __ __
20 15 19 5 20 7 19 2 20 16 7

__ __ __ __ __ __ __ __   __ __ __
25 15 13 14 19 22 25 4   2 6 25

__ __ __ __ __ __   __ __ __ __ __....
25 7 2 20 14 25   11 16 14 12 3

CRYPTOGRAM

Use the code to write the letters in the puzzle. The completed puzzle will reveal a quotation by William Faulkner.

A	B	C	D	E	G	H	I	J	K	L	M	N	O	P	R	S	T	U	V	W	Y
		14	1			12	24				5	15	23		3	16		4	6		

3 10 20 1'   3 10 20 1'   3 10 20 1'

10 6 10 3 9 22 12 24 15 2 —

22 3 20 16 12'   14 25 20 16 16 24 14 16'

2 23 23 1   20 15 1   26 20 1'   20 15 1

16 10 10   12 23 17   22 12 10 9   1 23

24 22.   7 4 16 22   25 24 13 10   20

14 20 3 18 10 15 22 10 3   17 12 23

17 23 3 13 16   20 16   20 15

20 18 18 3 10 15 22 24 14 10   20 15 1

16 22 4 1 24 10 16   22 12 10

5 20 16 2 10 3.   3 10 20 1!

ANTONYM SCRAMBLE

Write an antonym in the numbered spaces for each word. Then, transfer each numbered letter to the corresponding space in the grid. When you have completed the puzzle, you will find the answer to the following riddle: What did the man say when his friend saw snakes in his car?

1. girl — B (11) O (2) Y (28)
2. fair — ___ 13 ___ 3 ___ 23 ___ 24 ___ 33 ___ 16
3. alluring — ___ 39 ___ 30 ___ 37 ___ 19
4. prudent — ___ 7 ___ 10 ___ 44 ___ 34
5. freed — ___ 27 ___ 40 ___ 8 ___ 36 ___ 32
6. buck — ___ 1 ___ 6 ___ 22
7. calmness — ___ 29 ___ 12 ___ 21 ___ 43 ___ 20
8. verbose — ___ 41 ___ 15 ___ 4 ___ 18 ___ 9
9. neglect — ___ 17 ___ 42 ___ 31 ___ 38
10. straighten — ___ 26 ___ 5 ___ 35 ___ 25 ___ 14

1.	2. O	3.	4.	■	5.	6.	7.	8.
9.	■	10.	11. B	12.	13.	14.	■	15.
16.	■	17.	18.	19.	20.	21.	22.	■
23.	24.	25.	26.	■	27.	28. Y	■	29.
30.	31.	32.	33.	34.	35.	36.	37.	38.
■	39.	40.	41.	42.	43.	44.		

ANTONYM SCRAMBLE

Write an antonym in the numbered spaces for each word. Then, transfer each numbered letter to the corresponding space in the grid. When you have completed the puzzle, you will find the answer to the following riddle: How did the mathematician explain his weight to his colleagues?

1. clumsy ___ ___ ___ ___ ___ ___ ___ ___
   13 37 19 34 9 29 27 6

2. gentle ___ ___ ___ ___ ___ ___
   36 48 11 31 45 22

3. agitated ___ ___ ___ ___ ___ ___
   47 18 7 26 24 1

4. confining ___ ___ ___ ___ ___
   25 2 41 10 17

5. relevant ___ ___ ___ ___
   8 42 38 4

6. a common rock ___ ___ ___
   21 35 16

7. lengthen ___ ___ ___ ___
   40 20 12 39

8. single ___ ___ ___ ___ ___
   5 44 33 23 46

9. to refuse to hand out ___ ___ ___ ___
   43 32 15 30

1.	2.	3. **N**	4.	■	5.	6.	7.	8.
9.	■	10.	11.	■	■	12.	■	13.
14. **E**	15.	■	16.	17.	■	18.	19.	20.
21.	22.	■	23.	24.	25.	26.	27.	28. **M**
29.	30.	31.	32.	33.	34.	35.	■	36.
37.	38.	39.	■	40.	41.	42.	■	43.
44.	45.	46.	■	47.	48.			

LETTER DROP

1. Place the letters above each column into the correct spots in that same column. When you complete the puzzle, you will reveal a quotation by Albert Camus.

T M E S	U O R	N S	B T E	E D	N	H O W	T A I	T P	B P H	E Y	O	C T	W O H	N E E	C R
		■			■						'	■			■
				■				■			■				
					■					■					
	.														

2. Place the letters above each column into the correct spots in that same column. When you complete the puzzle, you will reveal a quotation by Aristotle.

W E E I	A E N A T	C T N	A E E	D R A	E L C	Y T T	H W	H D E	N A O B	U T	T	W E I	X S E A	C	N R E H	E L O A	L T B P
		■				■					■			■			
							■			.	■						
				'	■					'	■			■			
■			■				'	■				■		■			
		.															

LETTER DROP

1. Place the letters above each column into the correct spots in that same column. When you complete the puzzle, you will reveal a quotation by Johan Wolfgang von Goethe.

W	B	O	T	N	N	R	R	E	Y	E	N	M	I	Y	N
U	I	A	A	D	C	E	B	W	R	I	U	S	C	T	O
N	D	C	L	E	V	E	S	D	G	O	T		A	N	E
	M	O	S	I	O	P	O	S	E	H	A			A	D
	H	U	G				I	N		R	A			G	
		A								I					
								■				■			
■			,	■			■						■		
	■				,	■						■			.
■									■				■		
				,	■						,	■			
■						■			■			.			

2. Place the letters above each column into the correct spots in that same column. When you complete the puzzle, you will reveal a quotation by Thomas Jefferson.

I	E	E	C	A	S	W	A	R	T	P	I	T	N	E	T
H	T	T	O	U	S	R	E	F	S	W	R	I	N	S	T
Y	N	E	E	R	R	T	O	M	D		L	F	H	C	M
A	L	R	M	C	S	T	E	N			O	I	K		I
P	L				T		I	N				I			
A					K										
		■								■			■		
			,	■					■					■	
		■								;	■			■	
						■			■						
			,	■						■					■
	■					.									

LETTER DROP

Place the letters above each column into the correct spots in that same column. When you complete the puzzle, you will reveal a quotation by Franklin D. Roosevelt.

B	A	T	E	I	T	A	L	O	M	T	R	Y	O	D	N	M	L	S	T
D	N	G	I	I	T	C	O	M	T	H	R	R	N	E	O	U	N	T	N
I	O	V	R	K	E	L	A	T	M	O	I	T	S	S	B	A	E	A	A
O	R	T	A	Y	A		I	I	F	E	E	H		K	L	Y	E	D	H
I	T	Y	M	S			N			F	T	A		F	A	I	T		
T		D									N					S			
		■			■							■						■	
	■					■		■							■				■
			■			.	■			■			■						,
■						■			■								■		
	■				■								.	■				■	
				■				,	■				■						
			.																

ANSWER KEY

P. 3—Help Professor Whiz
1. Answers may vary but could include:

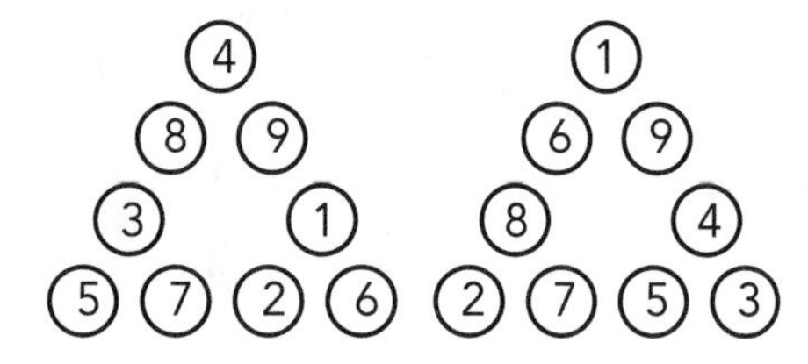

2. Answers may vary but could include:

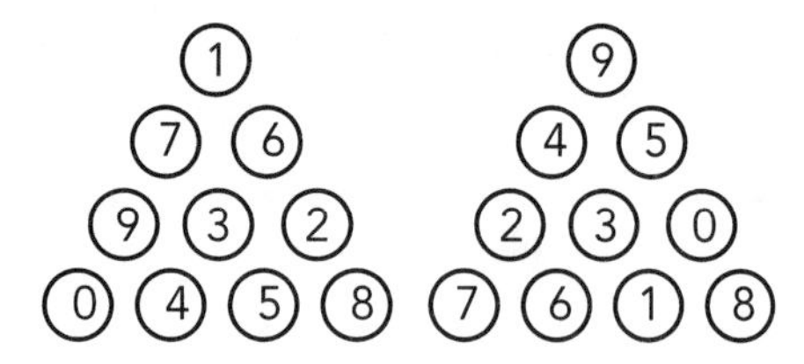

P. 4—Pyramid Dollars
1.

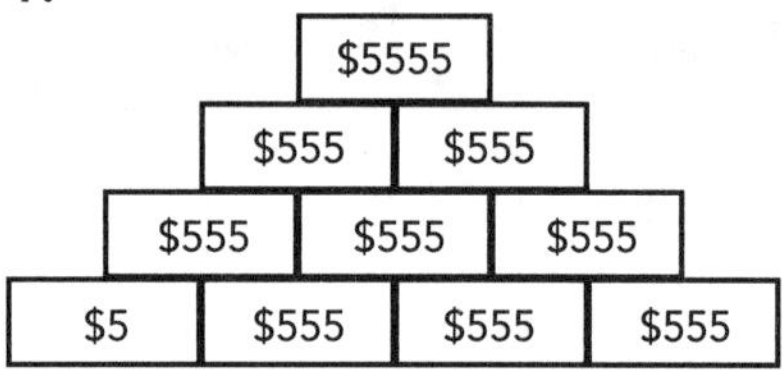

2.

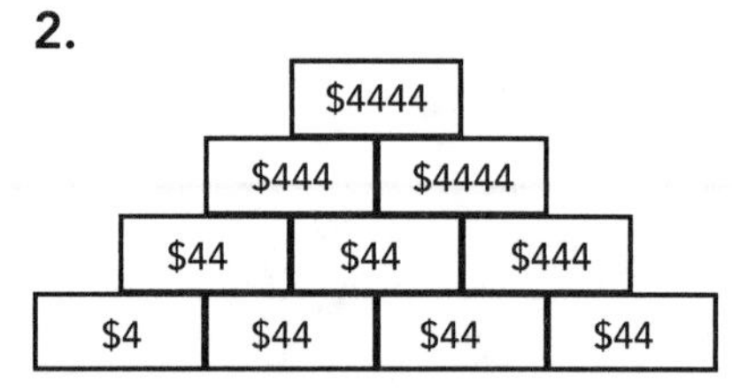

P. 5—Mystery Digits
1. 4; **2.** 2; **3.** 3; **4.** 6; **5.** 3; **6.** 9; **7.** 4; **8.** 7; **9.** 1; **10.** 5

P. 6—An Amazing Record
Step 2: 2,144,448,000 ÷ 60 = 35,740,800; **Step 3:** 35,740,800 ÷ 60 = 595,680; **Step 4:** 595,680 ÷ 24 = 24,820; **Step 5:** 24,820 ÷ 365 = 68; **Bonus!** The number 2,144,448,000 represents the

P. 6 (continued)
number of seconds that is equivalent to 68 years.

P. 7—Cousin Talk
1. 54 miles; **2.** Robert is 6' tall, and Gladys is 5'4" tall.; **3.** $163.84

P. 8—Mr. Kerr's Dogs
1. The dogs are 2, 3, and 6 years old. Since the product of the ages is 36, the possibilities are the following: 1, 3, 12; 2, 3, 6; 2, 2, 9; and 3, 3, 4. Because all of the dogs have been with Mr. Kerr at least two years, the ages 1, 3, and 12 are not correct. Because the sum of the numbers is an odd number, the ages cannot be 3, 3, and 4. Because there is an oldest dog and a youngest dog, the dogs must all be different ages.; **2.** German shepherd: Bandit, blue mat; Irish setter: Jericho, yellow mat; Scottish terrier: Pepper, red mat

P. 9—Ms. Baker's Brownies
Ms. Baker will need to bake two batches of brownies in the 12-inch pan to get 72 brownies that are the same size as the 16 brownies made from the 8-inch pan. To solve, first find the area of the 8-inch pan: 8" x 8" = 64 square inches. Divide 64 by 16 to find the area of each brownie: 4 square inches. Then, find the area of the 12-inch pan (12" x 12" = 144 square inches) and divide that by 4 square inches to find the number of brownies the pan will make (36). Finally, divide 72 by 36 to determine that 2 batches made in the 12-inch pan are needed to make the desired number of brownies.

P. 10—Multiplication Puzzler
1. 5; **2.** 5; **3.** 10,234; **4.** 24,987; **5.** 21,978

P. 11—Hot Dog Survey
80%

P. 12—At the Restaurant
1. 15; **2.**18; **3.** 6

P. 13—Line Them Up
1. Answers will vary. One possible solution is shown.

5	6	7	8
8	7	6	5
6	5	8	7
7	8	5	6

2. Answers will vary. One possible solution is shown.

5	6	7	8	9
7	8	9	5	6
9	5	6	7	8
6	7	8	9	5
8	9	5	6	7

ANSWER KEY

P. 14—Letter Grid

L	P	R
Q	M	T
U	S	N

P. 15—A Class Party
Ashley: grapes; Brad: cookies; David: punch; Emma: juice; Jill: cherries; Mike: popcorn; Rick: nuts

P. 16—Money Bags
The money is in the bags marked D, E, and H.

P. 17—At the Movies
The children are sitting as follows, from left to right and top to bottom: Anne: 14; Eleanor: 13; Martin: 17; Tom: 15; Lola: 16; Alex: 16

P. 18—Wacky Addition
1. b = 1, l = 7, o = 9, t = 8;
2. a = 0, d = 5, m = 1;
3. e = 2, i = 6, t = 1;
4. o = 0, t = 5, y = 1;
5. a = 8, l = 6, n = 4, s = l, t = 2;
6. a = 9, e = 3, m = 2

P. 19—Digit Patterns
2. 468, the sum of the first two digits add to the digit in the ones column; **3.** 4,137, the product of the first two digits is formed by the last two digits; **4.** 3,627, the sum of the two outer digits is equal to the sum

Page 19 (continued)
of the two inner digits; **5.** 8,156, the sum of the two outer digits is 1 less than the sum of the two inner digits; **6.** 4,090, the number formed by the first two digits added to the number formed by the last two digits equals 100

P. 20—A Pattern of Odd Numbers
Row 3: sum is 27; **Row 4:** 13, 15, 17, 19; sum is 64; **Row 5:** 21, 23, 25, 27, 29; sum is 125; **Row 6:** 31, 33, 35, 37, 39, 41; sum is 216; **1.** The sums are cubes.; **2.** 343; **3.** 1,000; **4.** Row 15; **5.** Row 20, **6.** Take the number of the row and find its cube.

P. 21—Dividing a Square
Yes, the square was divided into four equal parts. The student divided the square into four triangles. Each triangle has the same base (½ of a side of the square) and the same height (1 side of each triangle is equal to the side of the square). Since the area of a triangle is calculated by the formula ½bh, each triangle has the same area.

P. 22—Checkerboard Squares
There are 204 squares: 1 square is 8 x 8; 4 squares are 7 x 7; 9 squares are 6 x 6; 16 squares are 5 x 5; 25 squares are 4 x 4; 36 squares are 3 x 3; 49 squares are 2 x 2; 64 squares are 1 x 1; Formula: $n^2 + (n-1)^2 + (n-2)^2 + (n-3)^2 + (n-4)^2 + (n-5)^2 + (n-6)^2 + (n-7)^2$; n = 8

P. 23—Triangles and Circles
The left and right sides of the larger triangle are made of the diameters of the two circles. The sides of the smaller triangle are made of the radii of the two circles. See below.

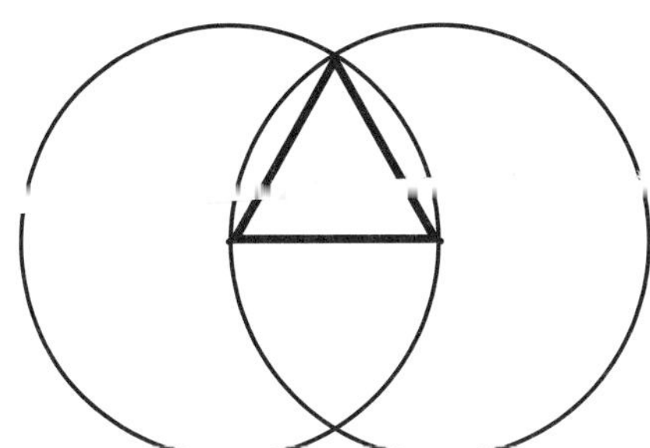

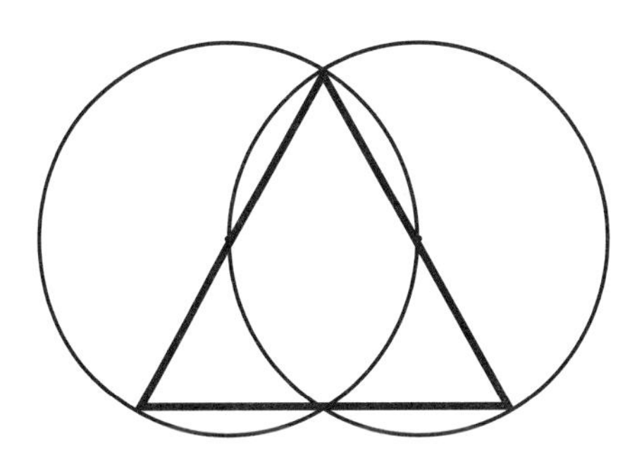

P. 24—Painters at Work
1. 1; **2.** 6; **3.** 12; **4.** 8; **5.** 8; **6.** 24; **7.** 24; **8.** 8

P. 25—Triangle Hunt
1. 8; **2.** 18; **3.** 56

P. 26—Connect the Dots

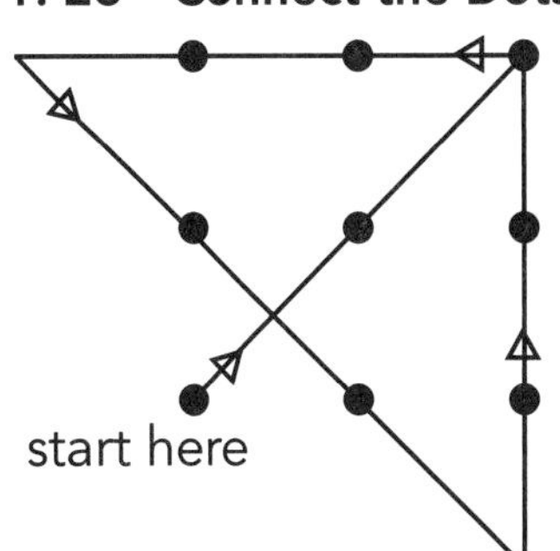

ANSWER KEY

P. 27—The Total

Jayla's fraction: $\frac{1}{4}$; Gabe's fraction: $\frac{5}{10}$; George's fraction: $\frac{4}{9}$; José's fraction: $\frac{7}{12}$; Maddie's fraction: $\frac{1}{5}$; Nathan's fraction: $\frac{1}{3}$; Paige's fraction: $\frac{3}{8}$; Wendy's fraction: $\frac{5}{6}$

P. 28—On Parade

Abby: snare drum; Sarah: trumpet; Chad: clarinet; Ian: bass drum; Jake: flute; Joseph: tuba; Olivia: bells; Valerie: saxophone; Zora: trombone

P. 29—Could Be the Answer

A. (4 + 5) x 8 – 2 x 4 = 64; **B.** [7 + (3 – 2)] x (3 + 2) – 2 x 8 = 24; **C.** (15 – 3) + 7 x (5 – 2) + 2 = 35; **D.** 8 ÷ [5 – (1 + 2)] + 5 x [9 – (3 + 4)] = 14; **E.** [(8 + 2) x 3 + 6)] ÷ 9; **F.** (8 + 2 + 5) ÷ 5 + (3 + 12) ÷ 3; **G.** (9 – 3) + (8 – 2 x 3); **H.** (7 – 1) x 3 – 10 ÷ (4 + 9 – 8); **I.** Answers will vary. One option: [(6 + 1) x 3 + 3] ÷ 6 = 4; **J.** Answers will vary. One option: (9 ÷ 3 + 5 x 4 + 1) ÷ 8 = 3; **K.** Answers will vary. One option: (4 x 5 + 2) – [5 x (12 – 8)] = 2; **L.** Answers will vary. One option: 9 x (6 + 5 – 8) = 27; **M.** Answers will vary. One option: [(½ + ¼) x ⅔ – ¼] + ⁶⁄₈ = 1; **N.** Answers will vary. One option: (2.4 – ½ ÷ ³⁄₆) – 6.6 x ⅙ = 0.3; **O.** Answers will vary. One option: 5.02 – (3 x ²⁄₆ + 0.4 x 3) = 2.82; **P.** Answers will vary. One option: (5 ¾ + 1.47) ÷ 0.04 ÷ (4 ⅚ – 2.4 ÷ 0.6) = 216.6

P. 30—Could Be the Answer

1. Answers will vary but may include: **A.** 6 + 4 x 2 ÷ 5 – 3 x 16 ÷ 2 + 6 = –10 ⅖; **B.** 6 + 4 x 2 ÷ (5 – 3) x 16 ÷ 2 + 6 = – 44; **C.** (6 + 4) x 2 ÷ [5 – 3 x 16 ÷ (2 + 6)] = -20; **D.** [(6 + 4) x 2 ÷ 5 – 3] x 16 ÷ 2 + 6 = 14; **E.** 6 + 4 x 2 ÷ (5 – 3) x 16 ÷ (2 + 6) = 14; **F.** (6 + 4) x 2 ÷ (5 – 3) x 16 ÷ (2 + 6) = 20; **2.** Answers will vary but may include: **G.** 9 – 3 x (2 + 1 + 2) x 12 ÷ 6 – 2 = -23; **H.** 9 – 3 x (2 + 1 + 2) x 12 ÷ (6 – 2) = -36; **I.** (9 – 3) x (2 + 1 + 2) x 12 ÷ (6 – 2) = 90; **J.** 9 – 3 x 2 + 1 + 2 x 12 ÷ 6 – 2 = 6; **K.** (9 – 3) x (2 + 1) + 2 x 12 ÷ 6 – 2 = 20; **L.** (9 – 3) x (2 + 1) + 2 x 12 ÷ (6 – 2) = 24; **3.** Answers will vary but may include: **M.** 24 ÷ 8 – 2 + 2 x 12 – 7 + 1 x 5 = 23; **N.** 24 ÷ (8 – 2) + 2 x 12 – 7 + 1 x 5 = 26; **O.** 24 ÷ (8 – 2) + 2 x (12 – 7) + 1 x 5 = 19; **P.** 24 ÷ [8 – (2 + 2)] x 12 – 7 + 1 x 5 = 70; **Q.** 24 ÷ [8 – (2 + 2)] x (12 – 7) + 1 x 5 = 35; **R.** 24 ÷ 8 – 2 + 2 x 12 – [(7 + 1) x 5] = -15

P. 31—Farmer's Market

corn	beans	X	celery
banana	orange	pear	broccoli
apricot	cauli-flower	plum	straw-berries
apple	tomato	grapes	carrot

P. 32—Puzzle Fill-In

Answers will vary.

P. 33—Steps

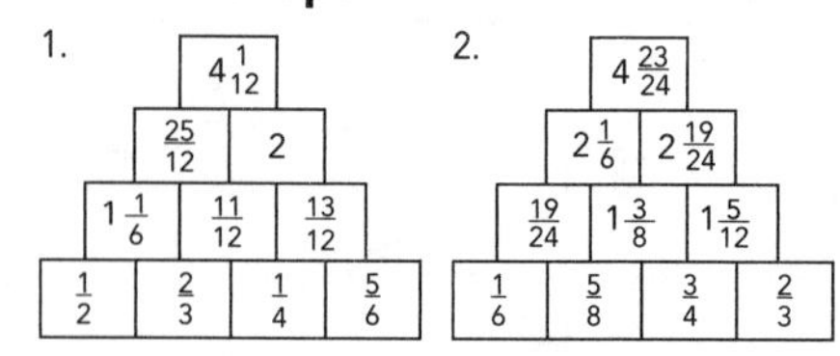

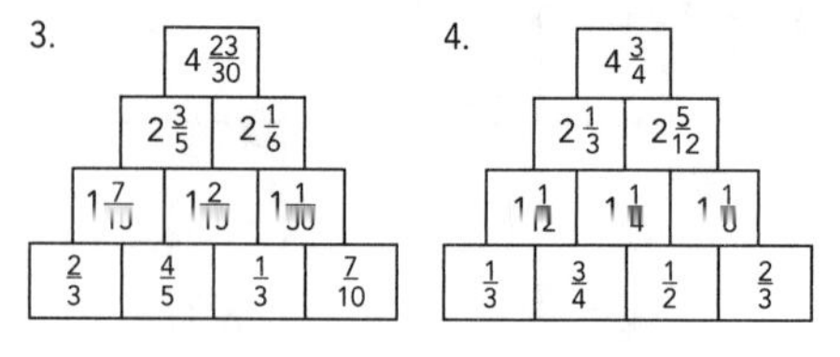

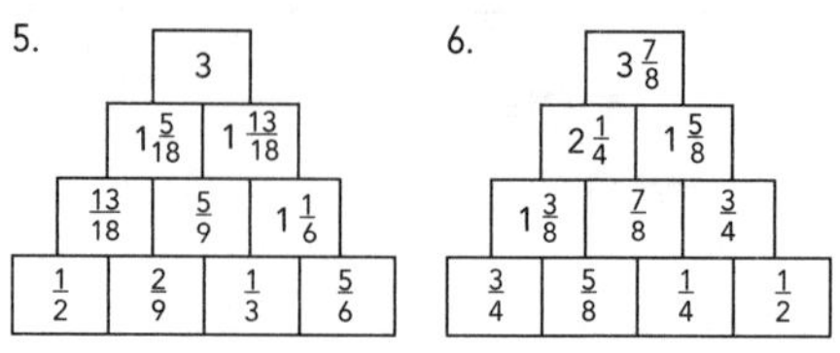

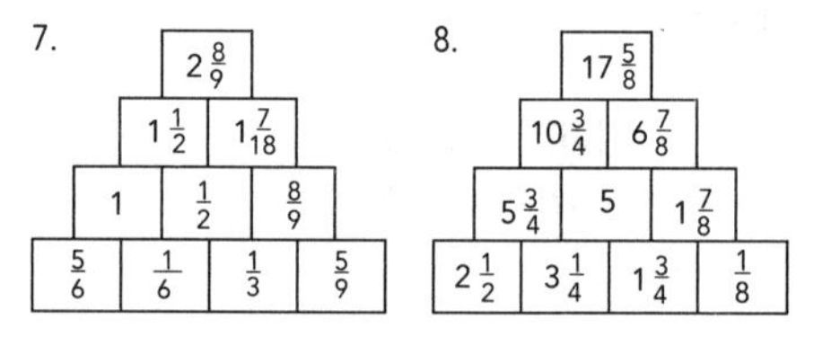

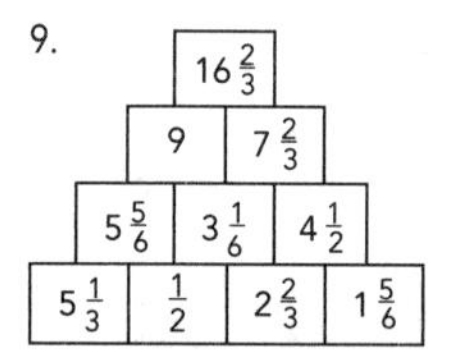

P. 34—Missing Steps

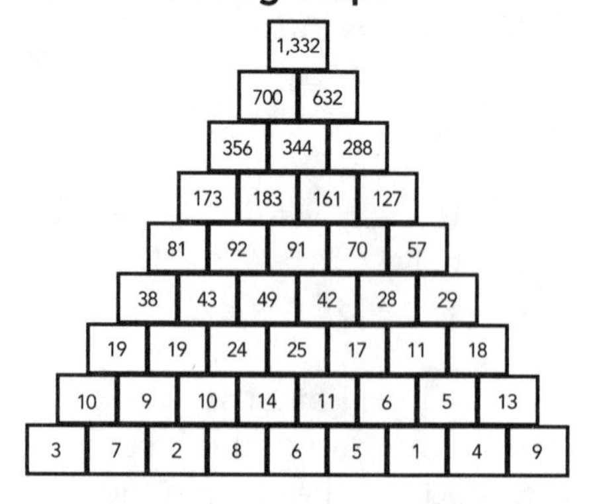

P. 35—Number Riddles

1. 2; **2.** -6; **3.** 12; **4.**-32; **5.** 7; **6.** 10; **7.** -21; **8.** -13; **9.** 1; **10.** -15; **11.** 5; **12.** -7

ANSWER KEY

P. 36—Money Mix-Up
Katie: $1.25; Derek: $2.88; Samuel: $3.53; Hailey: $4.50; Ian: $0.89; Jamison: $3.00; Melanie: $4.34; Ruth: $1.77; Trina: $2.61

P. 37—Word Fragments
1. plausible; **2.** abdicate; **3.** terrestrial; **4.** vicariously; **5.** blemish; **6.** chronological; **7.** gluttony; **8.** culmination; **9.** deliberate; **10.** extraneous; **11.** hierarchy; **12.** indigenous; **13.** judicious; **14.** zealous; **15.** lethargic; **16.** metamorphosis; **17.** nomadic; **18.** obsolete; **19.** orthodox; **20.** quandary; **21.** serendipity; **22.** submissive; **23.** utopia; **24.** juncture; **25.** whimsical; **26.** fortuitous

P. 38—Word Fragments
1. affluent; **2.** acclaim; **3.** judgment; **4.** alleviate; **5.** amorphous; **6.** cajole; **7.** chagrin; **8.** dexterity; **9.** aesthetic; **10.** exacerbate; **11.** fiasco; **12.** garrulous; **13.** harbinger; **14.** hybrid; **15.** impeccable; **16.** irascible; **17.** labyrinth; **18.** knickknack; **19.** mausoleum; **20.** sublimate; **21.** pandemonium; **22.** pauper; **23.** penultimate; **24.** pseudonym; **25.** queue; **26.** pamphlet

P. 39—Word Fragments
1. denounce; **2.** accentuate; **3.** behemoth; **4.** choreography; **5.** inarticulate; **6.** disheartened; **7.** embellish; **8.** flabbergasted; **9.** frugal; **10.** vocation; **11.** harrowing; **12.** hypothetical; **13.** abridge; **14.** jubilant; **15.** kudos; **16.** luminous; **17.** meticulous; **18.** ominous;

P. 39 (continued)
19. paramount; **20.** reputable; **21.** tranquil; **22.** coherent; **23.** quixotic; **24.** surreptitious; **25.** effervescent; **26.** plethora

P. 40—Common Links
1. jack; **2.** table; **3.** pick; **4.** smoke; **5.** water; **6.** cap; **7.** fork; **8.** play; **9.** check; **10.** note; **11.** show; **12.** day; **13.** night; **14.** moon; **15.** storm

P. 41—Rhyme Time
1. armchair; **2.** coffee table; **3.** bath towel; **4.** clothes dryer; **5.** ceiling fan; **6.** curtain rod; **7.** TV set; **8.** silverware; **9.** wallpaper; **10.** vanity mirror; **11.** bathtub; **12.** staircase; **13.** fax machine; **14.** water cooler; **15.** coffee maker; **16.** manila folder; **17.** snack machine; **18.** meeting room; **19.** corner office; **20.** laser printer

P. 42—Proverbs
1. haste makes waste; **2.** a fool and his money are soon parted; **3.** all roads lead to Rome; **4.** blood is thicker than water; **5.** don't cry over spilled milk; **6.** fit as a fiddle; **7.** his bark is worse than his bite; **8.** when it rains, it pours; **9.** make hay while the sun shines; **10.** in for a penny, in for a pound; **11.** never judge a book by its cover; **12.** no man is an island; **13.** no pain, no gain; **14.** a rolling stone gathers no moss; **15.** Rome was not built in a day

P. 43—Word Equations
1. 12 months in a year; **2.** 4 quarters in a dollar; **3.** 9 justices on the United States

P. 43 (continued)
Supreme Court; **4.** 29 days in February in a leap year; **5.** 24 karats in pure gold; **6.** 535 legislators in the United States Congress; **7.** 23 countries in North America; **8.** 5 toes on a human foot; **9.** 90 degrees in a right angle; **10.** 360 degrees in a circle; **11.** 7 wonders of the world; **12.** 18 holes on a golf course; **13.** 3 wheels on a tricycle; **14.** 52 cards in a deck, not counting the jokers; **15.** 1,760 yards in a mile; **16.** 8 planets in the solar system; **17.** 6 sides on a die; **18.** 100 decades in a millennium; **19.** 3 colors on a stoplight; **20.** 5 sides on a pentagon; **21.** 100 centimeters in a meter; **22.** 7 continents on Earth

P. 44—Analogies
1. D.; **2.** B.; **3.** B.; **4.** D.; **5.** A.; **6.** D.; **7.** C.; **8.** A.; **9.** C.; **10.** D.

P. 45—Book Bender
The books are in the following order: orange, green, teal, brown, purple, yellow, red, white.

P. 46—Car Conundrum
Brenda, Emily, and Gerald are riding in one car. Andy and Carl are riding in another car. Debbie and Frank are riding in the third car.

P. 47—Twisted Words
1. grammar; **2.** capitol; **3.** depicts; **4.** writing

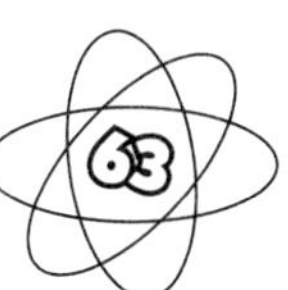

ANSWER KEY

P. 48—Think Links
1. cross boss; **2.** mad lad; **3.** keen teen; **4.** fair pair; **5.** real deal; **6.** brown gown; **7.** funny bunny; **8.** main pain; **9.** kind mind; **10.** dull gull; **11.–20.** Answers will vary, but may include: **11.** large lad; **12.** desolate dad; **13.** stinky stockings; **14.** grumpy grandpa; **15.** playful puppy; **16.** dirty digit; **17.** sad song; **18.** lost loot; **19.** small seaman; **20.** cheerful chum

P. 49—Beastly Words
Answers will vary but may include: **1.** swan; **2.** newt, tern, wren; **3.** elk; **4.** tuna; **5.** rat, toad; **6.** loon; **7.** seal; **8.** dog; **9.** gnu; **10.** mole, eel; **11.** bird; **12.** snake, crane, crake, saker; **13.** horse; **14.** goat; **15.** shark; **16.** deer; **17.** owl, owlet; **18.** wolf, owl, fowl; **19.** pig; **20.** frog

P. 50—That's "Pun-ny"
1. grouse; **2.** dandelion, **3.** tulips; **4.** yellow jacket; **5.** nightingale; **6.** caterpillar; **7.** thyme; **8.** lynx; **9.** swift; **10.** bedbug; **11.** carnation; **12.** yew; **13.** flycatcher; **14.** catnip; **15.** hyacinth

P. 51—Hidden Words
1. eye; **2.** head; **3.** knee; **4.** leg; **5.** arm; **6.** hand; **7.** skin; **8.** chin; **9.** seven; **10.** ten; **11.** zero; **12.** one; **13.** two; **14.** three; **15.** four; **16.** five

P. 52—Cryptogram
"Always bear in mind that your own resolution to succeed is more important than any one thing."

P. 53—Cryptogram
"Imagination is more important than knowledge. For knowledge is limited, whereas imagination embraces the entire world...."

P. 54—Cryptogram
"Read, read, read, everything—trash, classics, good and bad, and see how they do it. Just like a carpenter who works as an apprentice and studies the master. Read!"

P. 55—Antonym Scramble
2. unjust; **3.** vile; **4.** rash; **5.** mired; **6.** doe; **7.** worry; **8.** pithy; **9.** tend; **10.** twist; Don't worry about it; they're just my windshield vipers.

P. 56—Antonym Scramble
1. graceful; **2.** fierce; **3.** placid; **4.** roomy; **5.** moot; **6.** gem; **7.** trim; **8.** bunch; **9.** mete; Don't blame me, I get my large circumference from too much pi.

P. 57—Letter Drop
1. "To be happy, we must not be concerned with others."; **2.** "We are what we repeatedly do. Excellence, then, is not an act, but a habit."

P. 58—Letter Drop
1. "Whatever you can do, or think you can, begin it. Boldness has genius, power, and magic in it."; **2.** "In matters of style, swim with the current; in matters of principle, stand like a rock."

P. 59—Letter Drop
"It is common sense to take a method and try it. If it fails, admit it frankly and try another. But above all, try something."